Audio and Speech Processing with MATLAB®

Paul R. Hill

CRC Press
Taylor & Francis Group
Boca Raton London New York

CRC Press is an imprint of the
Taylor & Francis Group, an **informa** business

CRC Press
Taylor & Francis Group
6000 Broken Sound Parkway NW, Suite 300
Boca Raton, FL 33487-2742

Printed on acid-free paper
Version Date: 20181026

International Standard Book Number-13: 978-1-4987-6274-8 (Hardback)

Library of Congress Cataloging-in-Publication Data

Names: Hill, Paul (Researcher in image communication), author.
Title: Audio and speech processing with MATLAB / Paul Hill.
Description: First edition. | Boca Raton, FL : CRC Press/Taylor & Francis
Group, 2018. | Includes bibliographical references and index.
Identifiers: LCCN 2018027155| ISBN 9781498762748 (hardback : acid-free paper)
| ISBN 9780429444067 (ebook)
Subjects: LCSH: Speech processing systems. | Sound--Recording and
reproducing--Digital techniques.
Classification: LCC TK7882.S65 .H55 2018 | DDC 006.4/5--dc23
LC record available at https://lccn.loc.gov/2018027155

Visit the Taylor & Francis Web site at
http://www.taylorandfrancis.com

and the CRC Press Web site at
http://www.crcpress.com

To my Parents and Friends
Many thanks to my Dad for proofreading through the final
manuscript!

Contents

Preface

Audio and speech processing within the analogue and digital domains has a long history going back over a century to the origin of mechanical recording devices and the first days of telephony. Although contemporary speech and audio applications can trace their methods to these historical developments, DSP-based audio technologies have also necessarily established a unique set of algorithmic and mathematical tools within the subject. This advancement of audio-based DSP techniques and technologies has had the most profound effect on modern society. They have enabled the realisation of things previously thought of as science fiction, such as entire record collections being carried in the palm of your hand and voice recognition systems giving directions to the nearest café. The overall aim of this book is, therefore, to explore the background to speech and audio processing together with a comprehensive and up to date exploration of core aspects of the subject.

In order to understand all the divergent areas of speech and audio processing technology, an understanding of applicable physics, physiology, psychoacoustics, digital signal processing and pattern recognition is required. A comprehensive coverage of the applicable areas of these subjects is included in the first chapters of this book. This foundation is then used as the context of the later chapters that investigate diverse applications such as speech coding and recognition together with wideband audio coding. Many real-world example applications are also explored. Specifically, musical applications such as time stretching and recognition are examined in detail.

I have selected the topics carefully in order to reach the following audience:

> Students studying speech and audio courses within engineering and computer science departments.

> General readers with a background in science and/or engineering who want a comprehensive description of the subject together with details of modern audio standards and applications.

> Audio engineers and technically-based musicians who require an overview of contemporary audio standards and applications.

The text contains numerous real-world examples backed up by many MATLAB® functions and code snippets in order to illustrate the key topics within each chapter. The book and computer-based problems at the end of each chapter are also provided in order to give the reader an opportunity to consolidate their understanding of the content. Finally, starred sections identify text that is not key to understanding further sections, but can be read or returned to for interest to get an expanded understanding of the subject.

Although this book contains enough material to fill a two-semester graduate course, careful selection of material will enable it to be suitable for such a course, filling just one semester.

Dr. Paul Hill

Bristol, U.K.

List of Acroynms

AAC Advanced Audio Codec. 2, 5, 191

AbS Analysis by Synthesis. viii, 267, 271

ADPCM Adaptive Differential Pulse Code Modulation. 269

ADSR Attack, Decay, Sustain and Release. viii, 285, 291, 292

AIFC Audio Interchange File Format - Compressed. 2, 5

AIFF Audio Interchange File Format. 2, 5

AM Amplitude Modulation. 301

ASR Automatic Speech Recognition. vii, 195, 196, 198, 200, 202, 204

ATH Absolute Threshold of Hearing. 156

AU Sun Microsystems Audio File Format. 2, 5

BM Basiliar Membrane. 118

CD Compact Disc. 165, 166

CELP Code Excited Linear Prediction. 272

CELT Constrained Energy Lapped Transform. 281

CNNs Convolutional Neural Networks. 262

COLA Constant OverLap-Add. 88

CQF Conjugate Quadrature Filters. 170, 172

CTFT Continuous Time Fourier Transform. 59

DAW Digital Audio Workstation. 290

DNN Deep Neural Networks. 238, 263

DPCM Differential Pulse Code Modulation. 269

OGG OGG compression system or file. 2, 5

PCM Pulse Code Modulation. 269

PLP Perceptual Linear Prediction. 225

PQMF Pseudo-QMF FilterBank. 173

PR Perfect Reconstruction. 168

PSD Power Spectral Density. 80

QMF Quadrature Mirror Filters. 169

RM Ring Modulation. 304

RMS Root Mean Square. 36

RNNs Recurrant Neural Networks. 50, 238, 260

RoEx Rounded Exponential (filter). 135

RPE Regular Pulse Excited. 272

SHM Simple Harmonic Motion. 14, 16

SIL Sound Intensity Level. 151

SNR Signal to Noise Ratio. 36

SPL Sound Pressure Level. 151

STFT Short Term Fourier Transform. vi, 55, 83, 85

SVMs Support Vector Machines. 50

TIMIT Texas Instruments MIT Speech Dataset. 203

VCA Voltage Controlled Amplifier. 292

VCF Voltage Controlled Filter. 288, 292

VCO Voltage Controlled Oscilator. 288, 292

VST Virtual Studio Technology. 290

WOLA Weighted OverLap-Add. 86

Introduction

The perception of sound and its interpretation is a key human facility. It gives situational and spatial awareness, cues for visual perception and, most importantly, the ability to communicate. Communication can be in the form of sounds, music or, most importantly to this book, speech.

Audio and speech processing is, therefore, a unique and vitally important area of engineering. Within the context of applications such as Shazam, MP3 encoding and recognition systems such as Siri, speech and audio processing currently forms a key contemporary engineering research area; now and going forward into the future.

Overview

Figure 0.1 illustrates the overall structure of the subjects covered within this book. The left of this figure shows the analysis of physical audio signals. Applications corresponding to this type of audio processing include speech recognition systems (e.g., Siri), music identification (e.g., Shazam) and automatic music transcription. The right of this figure illustrates the synthesis of audio signals from digital representations. Applications corresponding to this type of processing include musical and speech synthesis.

Combining both the left and right sides of this figure illustrates coding applications that combine the analysis and synthesis of audio. These methods include the wideband coding of audio (e.g., MP3/AAC compression) and speech coding (e.g., CELP, Speex and Opus).

This book covers all three aspects of audio and speech processing illustrated in Figure 0.1. A small number of specific applications such as the phase vocoder for audio time dilation are included to illustrate the learnt techniques being applied in real world applications.

Learning Objectives

- Learn core engineering, mathematical and programming skills to process audio and speech signals

- Survey topics in sound analysis and processing

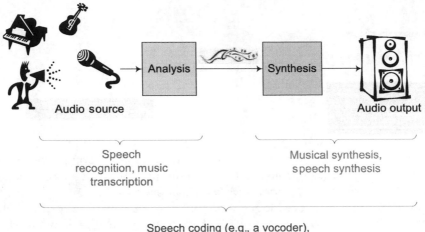

FIGURE 0.1: Illustration of various parts of the book

- Develop an intuition for sound signals

- Learn some specific technologies

- Be able to quickly manipulate and process audio with MATLAB

Book Features

As denoted by the title of this book, the programs contained are coded in MATLAB. Where possible, all the methods described within the book have accompanying code which attempts to explain or illustrate the main points of each technique. However, this is not possible for all of the material. Furthermore, there are some aspects of the book such as deep learning speech recognition that are effectively not possible to illustrate with MATLAB as they are so intricately linked with other languages and toolboxes (such as C/HTK or Python/Tensorflow). MATLAB code listings are shown enclosed in boxes. Command line inputs using MATLAB are also boxed. However, each line starts with the command line identifier >>.

This book can be read from cover to cover to give a good overview of the core mathematical, engineering and programming skills required for speech and audio processing. However, some contained elements can be considered to be more for the reader's interest and reference and are not key to understanding the overall content. These reference or background sections can be easily omitted on a first read and will not interfere with the ability of the reader to understand the rest of the book. These "optional" sections are labelled with a * symbol at the end of the sections' title.

Book Philosophy

This book dives straight into a detailed coverage of manipulating audio with MATLAB from the first chapter. This exemplifies the philosophy of its creation in as much as there is much more emphasis on practical insights into the subject rather than prosaic explanations around its background. A high-level treatise on audio as a human endeavour and attribute, etc. has been omitted in favour of a practical emphasis from the first page. It is hoped that this book will mostly inform but in some (possibly abstract) way entertain and inspire the reader. It has certainly been entertaining and inspiring to be totally immersed in the subject during its creation.

Notation

General

- Not equal to: $\neq$

- Equivalent to: $\equiv$

- Approximately equal to: $\approx$

- Proportional to: $\propto$

- Factorial of x: $x!$

- $\sqrt{-1}$: i or j

- x^T: Transform of x

Sets

- Natural Numbers: $\mathbb{N} = \{1, 2, 3, \ldots\}$

- Integers: $\mathbb{Z} = \{\ldots, -3, -2, -1, 0, 1, 2, 3, \ldots\}$

- Real Numbers: $\mathbb{R}$

- Positive Real Numbers: $\mathbb{R}_+$

- Complex Numbers: $\mathbb{C}$

- In a set: $\in$

- Range: (-1,1)

- Half open range: [-1,1)

Calculus

- First derivative: $\frac{\partial y}{\partial x} : y'$

- Second derivative: $\frac{\partial^2 y}{\partial x^2} : y''$

- Gradient: ∇y

Probability

- Probability: $P(w)$

- Conditional probability: $P(x|w)$

- Normal distribution: $\mathcal{N}$

MATLAB®

For product information, please contact:
The MathWorks, Inc.
3 Apple Hill Drive
Natick, MA 01760-2098 USA
Tel: 508-647-7000
Fax: 508-647-7001
E-mail: info@mathworks.com
Web: www.mathworks.com

FIGURE 0.2: Map of the structure of the book.

1

MATLAB® and Audio

CONTENTS

> To me, it's always a joy to create
> music no matter what it takes to
> actually get there. The real evils
> are always whatever stops you
> from doing that – like if your
> CPU is spiking and you have to sit
> there and bounce all your MIDI
> to audio. Now that's annoying!
>
> Skrillex

MATLAB (**MAT**rix **LAB**oratory) is used throughout this book for audio processing and manipulation together with associated visualisations. This chapter therefore gives an introduction to the basic capabilities of MATLAB for audio processing. Appendix B also gives a list of core MATLAB functions and commands that are applicable to audio processing for those starting to use the language. This chapter (and Appendix B) can be skipped or skimmed if the reader is familiar with the basic operations of the MATLAB programming language and its visualisation capabilities.

More general information is given on the Mathworks (creators of MATLAB) website `www.mathworks.com` and within the innumerable help files, demos and manuals packaged with MATLAB.

TABLE 1.1: Audio formats available to be read by MATLAB command
audioread

Audio File Format	Description	File extension
WAVE	Raw audio	.wav
OGG	OGG vorbis	.ogg
FLAC	Lossless audio compression	.flac
AU	Raw audio	.au
AIFF	Raw audio	.aiff,.aif
AIFC	Raw audio	.aifc
MP3	MPEG1 Layer 3, lossy compressed audio	.mp3
MPEG4 AAC	MPEG4, lossy compressed audio	.m4a, .mp4

1.1   Reading Sounds

Audio is read into MATLAB using the function audioread whose basic functionality is as follows. [1]

```
>> audioread(filename);
```

Where filename in this case is a MATLAB variable containing a string (array of chars) defining the entire name of the audio file to be read including any file extension (e.g., mp3, wav, etc.). A typical example of a call to audioread would be

```
>> [y Fs] = audioread('exampleAudio.wav');
```

where y is the array or matrix of sampled audio data and Fs is the sampling frequency of the input audio. audioread is able to read the formats shown in Table 1.1. In this example, filename is 'exampleAudio.wav', the file to be read in (filename is required to be of a MATLAB string type and is therefore delimited by single quotes '). filename can be a MATLAB string that can also include a path (defined in the format of your operating system) to any location on your hard drive. For example, filename could be 'c:\mydirectory\mysubdirectory\exampleAudio.wav' (on windows) or '~/mydirectory/mysubdirectory/exampleAudio.wav' (on OS-X/Unix/Linux). A statement in MATLAB will automatically display its results. It is therefore common to want to suppress this output and this is achieved by using the semicolon at the end of each line where no output is required.

It is often useful to determine detailed information about an audio file

---

[1]audioread replaces the more common wavread function for reading audio in previous versions of MATLAB. wavread has now been removed.

TABLE 1.2: Output structure of audioread

Structure element	Description
'Filename'	Filename
'CompressionMethod'	Description of compression method
'NumChannels'	Number of audio channels (either 1 or 2)
'SampleRate'	Number of samples per second
'TotalSamples'	Length of audio in samples
'Duration'	Length of audio in seconds
'Title'	
'Comment'	
'Artist'	
'BitsPerSample'	
'BitsRate'	

before (or indeed after) reading it using audioread. This is achieved using the audioinfo MATLAB function which has the following example usage:

```
>> information = audioinfo(filename);
```

where filename is defined as above (i.e., with audioread), and information is a MATLAB structure illustrated in Table 1.2. An example ouput is:

```
>> information = audioinfo('/Users/csprh/MUSIC/LocalMusic/02-
   Down-to-Earth.mp3')

information =

          Filename: '/Users/csprh/MUSIC/LocalMusic/02-Down-to-
             Earth.mp3'
 CompressionMethod: 'MP3'
       NumChannels: 2
        SampleRate: 44100
      TotalSamples: 14404608
          Duration: 326.6351
             Title: 'Down to Earth'
           Comment: []
            Artist: 'Prairie Dog'
           BitRate: 128
```

Once an audio sound has been loaded using audioread it can be played easily using the MATLAB functions sound or soundsc. [2]

---

[2]sound or soundsc are analogous to image and imagesc from the image processing toolbox.

1.2   Audio Display and Playback

A basic input (of a mono source) and display program using `audioread` is shown below:

```
1  filename = 'paulhill.wav'; %Define the name of the audio file
2  [Y, Fs]=audioread(filename); %Input the (mono) signal
3  timeAxis=(1:length(Y))/Fs; % Convert sample number to time
      vector
4  plot(timeAxis, Y);  % Plot the waveform
5  xlabel('time (seconds)');
6  title('Waveform');
7  sound(Y, Fs);   % Playback the sound
```

The output of this program is displayed in Figure 1.1.

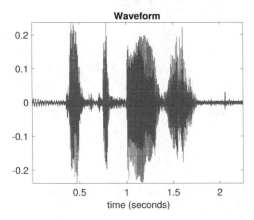

FIGURE 1.1: Basic display of a waveform using Matlab function `audioread`.

A subset of the audio samples can be read in using `audioread` using the following code:

```
1  filename = 'paulhill.wav'; %Define the name of the audio file
2  inputRange = 4000:4200;
3
4  [Y, Fs]=audioread(filename, [inputRange(1) inputRange(end)]); %
      Input the (mono) signal
5  timeAxis=inputRange/Fs; % Convert sample number ot time vector
6  plot(timeAxis, Y);  % Plot the waveform
7  xlabel('time (seconds)');
8  title('Waveform');
9  sound(Y, Fs);
```

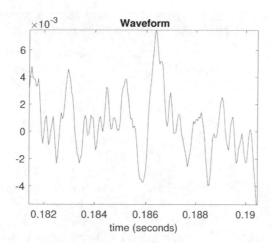

FIGURE 1.2: Basic display of a waveform subset using audioread.

The ouput of this program is shown in Figure 1.2.

1.3   Audio-Related MATLAB

Although wavread has been used in a large number of legacy systems, Mathworks has replaced it with the more flexible audioread function. The audioread function works in a very similar way to wavread but is able to input a much larger number of audio file types (WAVE (.wav), OGG (.ogg), FLAC (.flac), AIFF (.aiff, .aif), AU (.au) and AIFC (.aifc)). Within more recent operating systems MP3 (.mp3) and MPEG4 AAC (.m4a, mp4) are also supported as illustrated in Figure 1.1.

Conversely audiowrite is used to write audio data to a file. The following code shows how the data contained within one of the inbuilt audio .mat files in MATLAB can be written to a file.

```
1  load handel.mat;
2  %This inputs a variable Y (containing a mono audio signal) and
       the sampling
3  %frequency Fs
4  audiowrite('handel.wav',y,Fs);
```

A small list of undocumented example audio files included within MATLAB are shown in the box below. These can be used for fast prototyping and testing without the need to track down an example audio file. They are used extensively in the following example programs.

The format of the saved file is determined by the file ending of the file name

Example Audio Files Built into MATLAB for Testing

There are a small number of undocumented example audio files (speech, music and audio clips) included within MATLAB that can be easily used for testing.

File Name	Samples	Fs	Description
chirp.mat	13129	8192	Bird song
handel.mat	73113	8192	Sample of Handel's Messiah
mtlb.mat	4001	7418	Vocal sample of the word Matlab
train.mat	12880	8192	Train whistle
gong.mat	42028	8192	Gong being struck
laughter.mat	52634	8192	A number people laughing
splat.mat	10001	8192	Sound effect

An example of how to hear these files

```
load handel.mat
p = audioplayer(y, Fs);
play(p);
```

(i.e., in the above example the handel.wav file is a "wav" file). The supported formats are currently .flac, .m4a, .mp4, .oga, .ogg and .wav. Additionally, the data type of the output native data is determined by the data type of y in `audiowrite('handel.wav',y,Fs);`. A list of possible data types for each format (and other options) is given in the help file of `audiowrite`. [3]

Tables 1.3 and 1.4 show the main functions that are related to audio processing within MATLAB. More general MATLAB functions related to audio processing are described in Appendix B.

TABLE 1.3: Audio-Associated MATLAB Functions

Command	Description
sound(y)	play vector y as sound
sound(y, Fs, nBits)	play vector y as sound given sampling frequency Fs and number of bits nBits
soundsc(y)	scale vector y and play as sound

---

[3]Note: MP3 export is not supported by `audiowrite` (although it is supported by `audioread`). It must be assumed there are licensing issues surrounding this lack of support.

`soundsc(y, Fs, nBits)`	scale vector y and play as sound given sampling frequency Fs and number of bits nBits
`h = audioplayer(y, Fs);`	create an audioplayer object for vector y given sampling frequency Fs. A handle for the player is stored in h
`h = audioplayer(y, Fs, nBits);`	same as above, but with nBits number of bits
`h = audiorecorder(Fs, nBits, nChans);`	create and audiorecorder object with sampling frequency Fs, number of bits nBits and number of channels nChans
`record(h);`	start recording with audiorecorder/audioplayer object with handle h
`stop(h);`	stop recording with audiorecorder/audioplayer object with handle h
`pause(h);`	pause recording with audiorecorder/audioplayer object with handle h
`play(h);`	play audio loaded into audioplayer object or recorded into audiorecorder object (with handle h)
`p = audiodevinfo;`	output a list of input and output devices available into structure p

TABLE 1.4: Audio-Associated Signal Processing MATLAB Functions

Command	Description
`fft(x)`	perform an N point Fast Fourier Transform on array x where N is the length of x
`fft(x,N)`	perform an N-point Fast Fourier Transform on array x
`ifft(x)`	perform an N-point inverse Fast Fourier Transform on array x where N is the length of x
`ifft(x,N)`	perform an N-point inverse Fast Fourier Transform on array x
`fftshift(X)`	interchanges the output of the two halves of ffts output (useful for moving DC to index 1 in the array)
`[s,f,t]=spectrogram(x,w,n,f,Fs)`	create spectrogram of input vector x

`framedSignal = buffer(y, lenW, olap);`	create a set of overlapping windows (of length `lenW` and overlap `olap`) for buffer processing (such as window based energy evaluation)

1.4   Example Audio Manipulation

The following examples show some simple ways to manipulate sounds.

1.4.1   Reverse Audio in a File

To illustrate how to reverse an audio file the following code inputs an example audio sound (within the built-in .mat file handel.mat) and reverses the array of audio values using `flipud`. `flipud` is just a generic MATLAB function that reverses an array of numbers (flips them up and down in this case). [4]

```
1  load handel;
2  sound(y,Fs);
3  y2 = flipud(y);
4  sound(y2,Fs);
5  subplot(2,1,1);plot(y);
6  subplot(2,1,2);plot(y2);
```

Figure 1.3 shows the original and reversed audio signal from this code.

1.4.2   Downsample an Audio File

Downsampling by a factor of two is achieved with an audio file by discarding every other sample. An illustrative example implementation of downsampling by a factor of two is given below:

```
1  load handel;
2  sound(y,Fs);
3  y2 = downsample(y,2);
4  sound(y2,Fs);
5  subplot(2,1,2);plot(y2);
6  subplot(2,1,1);plot(y);
7  linkaxes;
```

The `downsample` function can be used to downsample by other (integer) values (greater than 1). The downsampled audio output is shown in Figure 1.4.

---

[4]If somehow you have transposed your audio data then you can use `fliplr`.

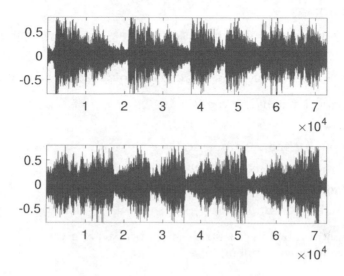

FIGURE 1.3: Reversed Audio: Original file on top, reversed below.

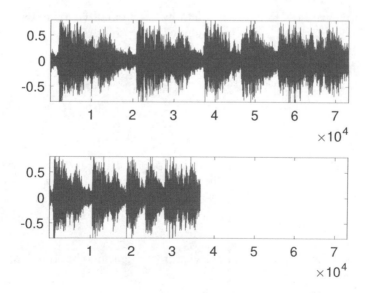

FIGURE 1.4: Downsampled Audio: Original file on top, downsampled below.

1.5 Summary

- This chapter introduces the basics of using the MATLAB programming language for audio processing.

- The basics of MATLAB are introduced including (also within Appendix B)

    - Basic functions and commands

    - Operators

    - Defining and initialising scalar, vector and matrix variables

    - Basic signal processing functions

    - Plotting functions

    - Audio related functions

- Example audio manipulations using MATLAB are given such as time reversal and downsampling

1.6 Exercises

In order to learn how to use MATLAB in its most effective form, try not to use loops when answering the following questions. The first questions should be answered in reference to the content in Appendix B.

Exercise 1.1

Assign a vector of length 50 so that all elements are of the value π to a variable a.

Exercise 1.2

Assign a vector of length n (where n is positive integer) to a variable b. b should contain a linearly increasing set of elements starting with 1, i.e., [1 2 3 4 ...].

Exercise 1.3

Assign a uniform random number vector (in the range [0,1]) of length 100 to a variable c.

Exercise 1.4

Assign a uniform random number vector (in the range [-10,10]) of length 1000 to a variable d.

Assign second and third variables e and f to arrays comprised of only the even and odd indicies of d respectively (i.e., e = [d(1) d(3) d(5) ...and f = [d(2) d(4) d(6) ...).

Exercise 1.5

Assign a uniform random number 2D matrix of size (100 × 100) to a variable A. Transpose a square Region of Interest (ROI) in the middle of A of dimension (10 × 10) starting with indices (45, 45).

Exercise 1.6

Create a square "magic" matrix of dimension (4 × 4), (5 × 5) and (7 × 7).

Exercise 1.7

Assign a vector of length 100 that contains a sine wave of output range [-100, 100] to a variable g. The array should contain exactly 10 cycles of the sine wave.

Hint: define an index range variable (as in Question 1.2) and manipulate this as the input to the MATLAB sin function.

Exercise 1.8

Download an example .wav file (there are several available in the code associated with this book). Produce a series of spectrograms (using MATLAB's built in spectrogram function). Change the windowing function from a Hanning to a Hamming window and note the difference.

2

Core Concepts

CONTENTS

> But in my opinion, all things in
> nature occur mathematically.
>
> René Descartes

2.1 Introduction

This chapter introduces some of the basic concepts necessary to understand the techniques and methods described within the remainder of this book.

The fundamental concepts of frequency analysis within the context of audio and speech processing will be covered in considerable detail in the next chapter. However, these concepts will need a detailed understanding of complex numbers as they are applied within signal processing. This is because nearly all forms of filtering and frequency analysis/processing use mathematical constructs based on complex numbers. This chapter includes a detailed

14

introduction to complex numbers and their utility in representing sampled signals. Initially, however, the fundamental nature of physical vibrations together with simple harmonic motion and sinusoids are first explored. Finally, this chapter gives an introduction to the Z transform and machine learning, core concepts necessary to understand digital filtering and automatic speech recognition, respectively.

2.2 Natural Vibrations and Sinusoids

Due to both the nature of many physical processes and their compact and easy representation with trigonometric functions, sinusoids are a key building block for the analysis and representation of audio signals.

Simple Harmonic Motion (SHM) is the most basic and most easily understood (mathematically) periodic oscillation. From simple harmonic motion, we derive the wave equation in the next chapter firstly for a vibrating string and then for a vibrating column of air. These are fundamental building blocks for the analysis of musical and speech audio signals and in fact audio signals in general. For example, within subsequent chapters, a model for the formant frequencies of speech is derived from the wave equation for a vibrating column of air.

2.2.1 Simple Harmonic Motion (SHM)

SHM is based on equating Hooke's Law to Newton's Second Law.

Hooke's Law

Hooke's Law states that a force is proportional to the displacement of a physical system (for example, a stretched string). This linear relationship can be defined as:

$$F \propto -x, \tag{2.1}$$

where F is the force (in Newtons) and x is the displacement, i.e., the force of a displacement is proportional to that displacement (in the opposite direction). Figure 2.1 shows a typical relationship between F and x over all possible values of x for a hypothetical material. The region of proportionality at the bottom left of the graph shows the region where Hooke's Law holds true. Hooke's Law is true for most metals (and many other materials) up to just below what is known as the "yield point" also shown in this figure.

A constant of proportionality can be defined as k. Hooke's Law can then be defined as:

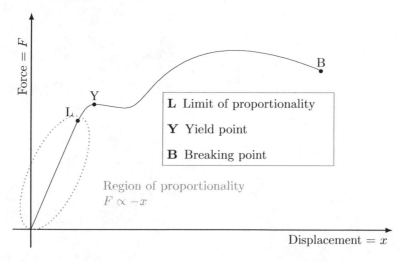

FIGURE 2.1: A graph showing forces versus displacement. The linear section illustrates Hooke's Law.

$$F = -kx, \tag{2.2}$$

where k has units Newtons per metre (N/m) and k is a positive constant representing the relative restoring force within the system.

Newton's Second Law

Newton's second law states that force F (measured in Newtons) equals the mass of an object m multiplied by the acceleration of the object a, i.e.,

$$F = ma, \tag{2.3}$$

or equivalently (as acceleration is the second derivative of displacement x with respect to time t):

$$F = m\frac{\partial^2 x}{\partial t^2}. \tag{2.4}$$

Simple Harmonic Motion Equation and its Solution

Two abstract SHM scenarios are shown in Figure 2.2. The left subfigure shows the vibration of a tine of a tuning fork. In this scenario, the displacement x is the distance from the central position of the tine and F is the returning force of the "elastic" nature of the metal tine. In the right-hand subfigure, a

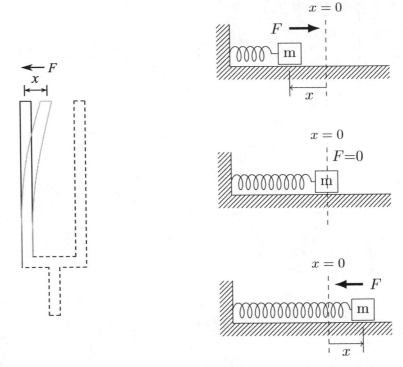

(a) Simple harmonic motion example 1: The vibration of a tuning fork tine after being struck
by a hammer.

(b) Simple harmonic motion example 2: The
motion of a mass attached to a spring. No
friction is assumed (i.e., it is undamped).

FIGURE 2.2: Examples of SHM.

weight is attached to a horizontal spring and vibrates (without friction) about
a central position. In this scenario, the force F is the returning force of the
spring and x is the displacement distance from the central position.

The equation of SHM is formed by equating the Force term F of (2.4) and
Hooke's Law (2.2), i.e.,

$$m\frac{\partial^2 x}{\partial t^2} = -kx. \tag{2.5}$$

A solution to (2.5) is a simple sinusoid given in the form:

$$x = A\sin(\omega t + \phi), \tag{2.6}$$

where A, ω and ϕ are the amplitude, angular frequency and phase of the
sinusoid, respectively. This can be proved to be a solution to (2.5) by replacing

x in (2.5) by x defined within (2.6) as follows. Firstly we differentiate (2.6) with respect to t twice:

$$\frac{\partial}{\partial t} A\sin(\omega t + \phi) \quad = \quad \omega A\cos(\omega t + \phi), \tag{2.7}$$

$$\frac{\partial^2}{\partial t^2} A\sin(\omega t + \phi) \quad = \quad -A\omega^2\sin(\omega t + \phi). \tag{2.8}$$

Equating (2.5) and (2.8) it can be seen that (2.6) is a solution to (2.5) given that:

$$\omega = \sqrt{\frac{k}{m}}. \tag{2.9}$$

This is intuitively correct as when mass increases in a vibrating system, the frequency decreases. Conversely, as the Hooke's Law proportionality constant increases (e.g., the "stiffness" of a vibrating string) the frequency will increase. For example, guitar strings with heavier mass will be lower in pitch and the more they are tightened the higher their pitch.

Examples of simple harmonic motion include [1]:

- The air pressure variation at a point generated by a single audio tone

- The string of a guitar after it has been plucked

- The pendulum of a pendulum clock

- The Alternating Current (AC) found in common electrical circuits

2.2.2 Sinusoid Vibrations

The above analysis shows that sinusoids are the solution to a broad range of physical problems. Their ubiquity can in some regards be considered to be a mathematical convenience. However, it is not unreasonable to assume (given the above analysis) that they are fundamental to any understanding of signal processing and, by extension, audio processing.

Formal Definition of a Sinusoid for Audio Signal Modelling

A real valued sinusoid of infinite extent (illustrated in Figure 2.3) can be defined as:

$$x(t) = A\cos(2\pi f_0 t + \phi), \tag{2.10}$$

where

---

[1]Given suitable simplifying physical assumptions.

- $x \in \mathbb{R}$ is the real function output

- $A \in \mathbb{R}_+$ is the amplitude of the sinusoid defined in the range: $[0, \infty)$

- $t \in \mathbb{R}$ is the real valued and infinite extent time index defined in the range: $(-\infty, \infty)$

- ϕ is the phase of the signal in the range: $(-\pi, \pi]$

- f_0 is the frequency of the signal defined in the range: $(-\infty, \infty)$.

If t is defined in seconds, then f_0 is defined in Hertz (defined as cycles per second). In many cases, the angular (or radian) frequency ω_0 is more useful within trigonometric and related mathematical functions. It is defined as:

$$\omega_0 = 2\pi f_0. \tag{2.11}$$

It should be noted that although negative frequencies make no intuitive sense, they are a useful construct. [2]

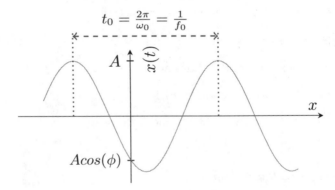

FIGURE 2.3: Sinusoid defined by $x(t) = A\cos(2\pi f_0 t + \phi)$.

2.3 Complex Numbers for Signal Processing

Complex arithmetic centres on the fundamental definition of quantity i as being the square root of -1, i.e., $i \equiv \sqrt{-1}$ (see Box 2.3.1). A key further development in complex arithmetic was the combination of a purely imaginary

---

[2]A good illustration of a negative frequency would be a situation where the frequency of a rotating wheel is measured. A negative frequency in such a scenario can be considered to be just the wheel rotating in the opposite direction.

term with a real term to form a general complex number and especially how this then represents points within the Euclidean plane.

Box 2.3.1: The definition of i

The definition of i (also alternatively defined as j) is a value that is defined as the square root of -1, i.e., $i \equiv \sqrt{-1}$. Although this is the universally accepted definition, I believe that i is better defined as a quantity that when squared equals -1, i.e., $i^2 \equiv -1$. I believe this is a more intuitive definition and removes the confusion surrounding the possible multi-values of a square root. The ubiquity of i within signal (and audio) processing is derived from the unique behaviour when multiplying two complex numbers (see Section 2.4.3). This behaviour is explicitly dependant on the fact that when i is multiplied by itself -1 is the result. Although often hard to grasp initially, as long as you can accept that there exists a defined value i that when it is multiplied by itself, the result is -1, then all of the utility of the subsequently defined mathematical behaviour follows (see Section 2.4 and Chapter 3).

Complex numbers therefore are now represented by convention in the form $x + iy$ where x and y are scalar values that denote the real ($\Re$) and imaginary ($\Im$) components, respectively. Complex numbers in this form can be represented on an **Argand Diagram**. An **Argand Diagram** is similar to the **Cartesian Coordinate System** except that the imaginary axis and real axis replace the y and x axes, respectively (which you would commonly expect to see on the Cartesian plane). An example of a complex value $x + iy$ within an Argand diagram is shown in Figure 2.4.

The development of complex numbers has, in itself, had a complex history starting in the 16th century as explained in appendix A. The now commonly accepted form of complex numbers $x + iy$ and its geometrical representation was proposed by Caspar Wessel and further developed by Jean-Robert Argand whose name bears the name of these geometric diagrams (see Figure 2.4).

This spatial representation is especially useful in understanding the basic operations on complex numbers. Specifically, the addition and multiplication of complex numbers follow the laws of vector addition (Figure 2.5) and rotation/expansion explained below (Figure 2.6).

Within MATLAB, all single numbers and the elements of vectors or matrices can be either real or complex. A simple example of a variable A being initialised as a complex number is (as illustrated in Figure 2.4)

```
>> A = 2.2 + 2.5*i;
A =
   2.2000 + 2.5000i
```

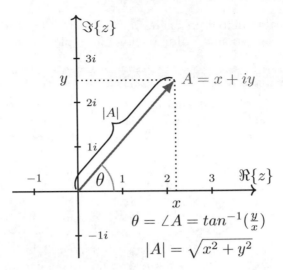

FIGURE 2.4: Example of an Argand diagram with $x = 2.2$ and $y = 2.5$: $A = 2.2 + 2.5i$.

The length of a complex number "vector" is known as its magnitude or absolute value (and more rarely its modulus). It is defined mathematically as $|A|$ and is calculated from Pythagoras' theorem as $|A| = \sqrt{x^2 + y^2}$ where $A = x + iy$. In MATLAB, complex magnitude is obtained using the abs function:

```
>> A = 2.2 + 2.5*i;
>> abs(A)
ans =
   3.3302
```

The angle of a complex number (the angle measured anticlockwise from the real axis to the complex vector) is known as its argument (or just angle) and is defined as $\angle A = tan^{-1}(y/x)$ where $A = x + iy$. [3] In MATLAB, a complex value's angle/argument is obtained using the angle function, e.g.,

```
>> A = 2.2 + 2.5*i;
>> angle(A)
ans =
   0.8491
>> radtodeg(angle(A))
ans =
   48.6522
```

---

[3]There are complexities in the calculation of a complex number's argument due to the undefined nature of (y/x) when $x = 0$. A specific function known as atan2 is often defined and used to calculate the argument within many programming languages; however, within MATLAB we just use angle for a complex number (see above).

The answer to `angle` is always given in radians. This is converted to degrees in the above listings using the MATLAB function `radtodeg`.

2.4 Simple Operations with Complex Numbers

2.4.1 Addition

Addition of complex numbers is defined as the separate addition of the real and imaginary parts of the added complex numbers.

$$(a + ib) + (c + id) = (a + c) + i(b + d) \tag{2.12}$$

With the geometric representation of complex numbers (where each number is represented by a vector on an Argand diagram), complex number addition is represented by vector addition as illustrated geometrically in Figure 2.5.

Within MATLAB, as all single numbers can be either real or complex the addition (or subtraction) operation of two complex numbers is identical to two real numbers. A simple addition example of two complex variables (A and B) can be represented within MATLAB as:

```
>> B = 1 + 3*i;
>> A = 3 + i;
>> C = A + B
C =
   4.0000 + 4.0000i
```

2.4.2 Subtraction

Subtraction of complex numbers is defined similarly to addition, i.e., as the separate subtraction of the real and imaginary parts of the subtracted complex numbers:

$$(a + ib) - (c + id) = (a - c) + i(b - d). \tag{2.13}$$

2.4.3 Multiplication

Multiplication of two complex numbers is defined as:

$$(a + ib)(c + id) = ac + ibc + iad + bdi^2 = (ac - bd) + i(bc + ad). \tag{2.14}$$

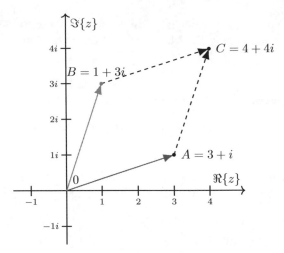

FIGURE 2.5: Addition of two complex numbers.

The unique mathematical and geometrical behaviour associated with the multiplication of two complex numbers is the main reason they have found fundamental utility within the worlds of geometry and signal processing (and therefore audio signal analysis/processing). The multiplication of two complex numbers is illustrated in Figure 2.6. Within this figure the top left sub-figure shows the two complex numbers (A and B to be multiplied and their product C). The top right sub-figure shows one of the two complex numbers B and its product with i, iB. Through visual inspection, the multiplication of a complex number by i can be directly shown to incur an anti-clockwise rotation of a complex vector by 90 degrees. Therefore iB is at right angles to B but has the same length. The bottom figure shows that the product of two complex numbers can be represented by a real number multiple of B added to a real number multiple of iB. This figure directly illustrates that angles of the two complex numbers A and B are added when they are multiplied (giving their complex product (C)), i.e., $\angle C = \angle A + \angle B$, $\theta = \phi + \psi$.

2.4.4   Euler's Formula

Euler's identity (2.15) gives an extremely useful alternative representation of a complex number:

$$re^{i\theta} = r\cos(\theta) + ri\sin(\theta) \tag{2.15}$$

Where r is the magnitude of the complex number and θ is its argument. A visual representaiton of such a complex number is shown in Figure 2.4, where

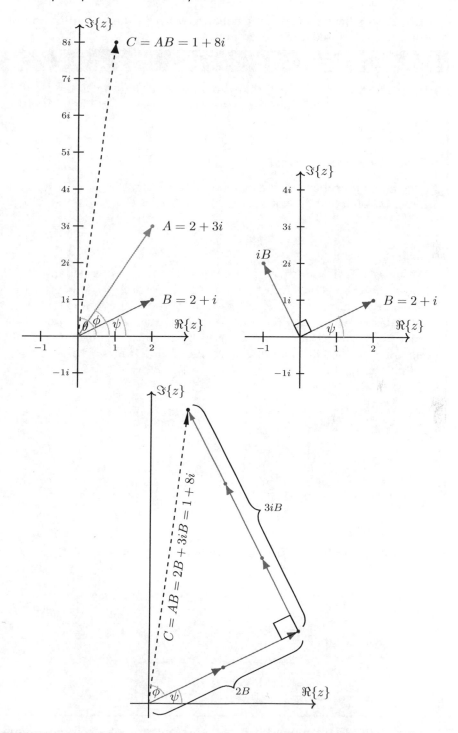

FIGURE 2.6: Multiplication of two complex numbers A and B showing their lengths are also multiplied and their angles are added.

$r = |A|$. Euler's formula (the same equation without the multiplication of a magnitude) is explored and proved in Box 2.4.1.

Box 2.4.1: Euler's Formula (Proof)

Euler's formula (see below) is key to understanding the use and utility of complex numbers for the analysis of discrete and continuous signals.

Proof by Series Expansion

The defining property of the exponential function $f(x) = e^x$ is that it equals its own first derivative, i.e.,

$$\frac{d}{dx} f(x) = f(x). \tag{2.16}$$

Although a Taylor series representation of e^x can be generated in other ways, the expansion given by (2.17) can be proven to satisfy (2.16) through the term by term derivative calculation. Doing this term by term derivative calculation leads to the same sequence (The polynomial order of each term is decreased by one: The series expansion is shifted to the left but still equals the original function due to the infinite sum).

$$f(x) = e^x = 1 + \frac{x}{1} + \frac{x^2}{2!} + \frac{x^3}{3!} + \dots \tag{2.17}$$

$$e^{i\theta} = 1 + \frac{i\theta}{1} + \frac{(i\theta)^2}{2!} + \frac{(i\theta)^3}{3!} + \dots \tag{2.18}$$

$$= 1 + \frac{i\theta}{1} - \frac{\theta^2}{2!} - \frac{i\theta^3}{3!} + \dots \tag{2.19}$$

Given that the Taylor series expansions of sin and cos functions are given by:

$$\cos(\theta) = 1 - \frac{\theta^2}{2!} + \frac{(\theta)^2}{4!} - \dots \quad and \quad \sin(\theta) = \theta - \frac{\theta^3}{3!} + \frac{\theta^5}{5!} - \dots \tag{2.20}$$

It is easy to see that

$$e^{i\theta} = \cos(\theta) + i\sin(\theta) \tag{2.21}$$

Equation (2.15) is in effect a polar representation of a complex number and can be used to confirm the result that the multiplication of two complex numbers is just the multiplication of their magnitudes and the addition of their arguments.

$$r_3 e^{i\theta} = r_1 e^{i\phi} r_2 e^{i\psi} = r_1 r_2 e^{i(\phi+\psi)} \tag{2.22}$$
$$r_3 = r_1 r_2$$
$$\theta = \phi + \psi$$

A straightforward proof of Euler's formula is given in Box 2.4.1. This Box also gives an expansion of the polar exponential form that is also visualised in Figure 2.7.

2.4.5   Visualisation of Euler's Formula*

Euler's formula equates a complex exponential to a complex sinusoid. Further intuition of this remarkable result is shown in Figures 2.7 and 2.8. Firstly Figure 2.7 geometrically visualises the expansion of $Ae^{i\theta}$ given in (2.18). As each term in this expansion is itself a vector it can be shown that the expansion always geometrically converges to the point $Ae^{i\theta}$.

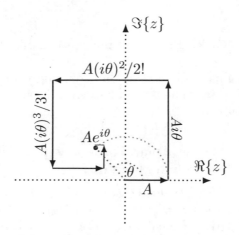

FIGURE 2.7: Visualisation of $Ae^{i\theta}$ using (2.18).

2.4.6   Visualisation of a Complex Phasor*

The representation of a sinusoid given in (2.10) can be generalised to a complex sinusoid known as a phasor (utilising (2.15)):

$$x(t) = Ae^{i(\omega_0 t + \phi)} = A\cos(\omega_0 t + \phi) + iA\sin(\omega_0 t + \phi). \tag{2.23}$$

This shows that a complex vibration can be simply represented using a complex exponential. This is a vital representation utilised by the Fourier analysis techniques described in detail in the next chapter.

Disregarding the amplitude A and the phase ϕ, a rotating unit phasor is illustrated in Figure 2.8. This figure shows that as the complex phasor rotates it traces out a cosine function on the real ($\Re$) axis and a sine function on the imaginary ($\Im$) axis.

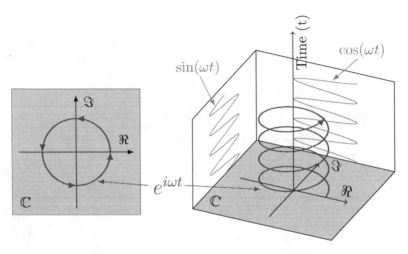

FIGURE 2.8: The components of a phasor.

The real and imaginary components of a rotating complex phasor can be separated into two sinusoids illustrated using the following MATLAB code (illustrated by Figure 2.9):

2.4.7 Conversion from Polar Form Complex Sinusoid to Real Sinusoid: A Key Euler Identity

Expressions for simple real valued cosine- and sine-based sinusoids given in (2.24) and (2.26) are easily derived from (2.19). Figure 2.10 illustrates a key Euler identity (2.25) derived from (2.24). This identity shows how a real-valued sinusoid can be extracted from the polar (exponential) form of complex sinusoids. The identity (2.25) shows that a real-valued cosine-based sinusoid can be extracted from the polar form $e^{-i\omega t}$ by averaging itself and its conjugate. Equations (2.25) and (2.27) show how to obtain both a cosine-based sinusoid and a sine-based sinusoid from the polar form.

Equation (2.25) illustrates an important result in that it shows that any real signal is symmetric in the frequency domain.

Listing 2.1: Code to separate complex exponential into sinusoids

```
1   A = 1;
2   N = 32;
3   fs = N/(2*pi); % Sampling frequency (samples/sec)
4   t = -3*pi:1/fs:3*pi; % Time Vector
5   omega = 1;
6   x = A*exp(i*omega * t);
7   plot(t,real(x),'-'); hold on;
8   plot(t,imag(x),'-.');
```

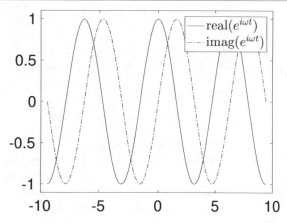

FIGURE 2.9: Illustration of a rotating phasor separated into its real and imaginary components using MATLAB code in Listing 2.1.

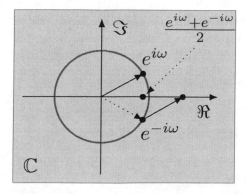

FIGURE 2.10: Visualisation of key Euler identity (2.25).

$$\cos(\omega t) = \frac{e^{i\omega t} + e^{-i\omega t}}{2} \tag{2.24}$$

$$= \frac{e^{i\omega t}}{2} + \frac{e^{-i\omega t}}{2} \tag{2.25}$$

$$\sin(\omega t) = \frac{e^{i\omega t} - e^{-i\omega t}}{2i} \tag{2.26}$$

$$= i\frac{e^{-i\omega t}}{2} - i\frac{e^{i\omega t}}{2} \tag{2.27}$$

2.5 Discrete Time Signals

2.5.1 Introduction

Given a contiguous sequence of integer indices $\{n\}$ (where $n \in \mathbb{Z}$), a discrete time signal can be defined as $\{x[n]\}$ where $x[n]$ is the nth sample ($x[n] \in \mathbb{R}$). Given that n is in the range $n_1 \le n \le n_2$ the discrete time signal in this indexed range can be represented as $\{x[n]\}_{n_1}^{n_2}$. Generic discrete signals are synonymous with discrete time signals. However, the indexing variable n can be defined in any domain (such as time or space). For audio signals, such discrete representations are indeed most accurately described as Discrete Time Signals and the indexing variable n is synonymous with time.

It should be clear that the use of square brackets in terms such as $x[n]$ denotes a discrete signal. A continuous signal $x(x)$ (usually denoted by round brackets) can be sampled to form a corresponding discrete signal $x[n]$. In such a case there is an associated sampling interval time T between two successive samples. The related quantity F_s (defined as $F_s = \frac{1}{T}$) is known as the sampling frequency. The sampling instances of the sequence are therefore at $t = nT$. It should be noted that:

- for real discrete time signals: $x \colon \mathbb{Z} \to \mathbb{R}$

- for complex discrete time signals: $x \colon \mathbb{Z} \to \mathbb{C}$

- $x[n]$ is only defined for integer values of n

- $x[n]$ can be represented graphically using a stem plot (e.g., using the MATLAB stem function)

- $x[n]$ is the nth sample of the signal

2.5.2 Finite Length Signals

A finite length signal is a discrete-time signal of finite length and is a contiguous list of N real values. A finite length discrete-time signal can be directly represented as a vector in $\mathbb{R}^N$. Such a vector can be represented in many forms. It can be defined in standard vector notation as:

$$x = [x_0, x_1, \ldots, x_{N-1}]^T.$$ (2.28)

Alternatively, the sequence can be represented by its sequenced indices n:

$$x[n], \quad n = 0, 1, \ldots, N - 1.$$ (2.29)

MATLAB Finite Length Signal Representations
A finite duration discrete signal can be represented in MATLAB using either a column or row vector. An example discrete time sequence $\{x[n]\} = \{1, 5, 8, 9, -4, 3, 2\}$ can be represented in MATLAB as

```
>> x = [1,5,8,9,-4,3,2];
```

This can be visualised using the MATLAB `stem` function producing Figure 2.11.

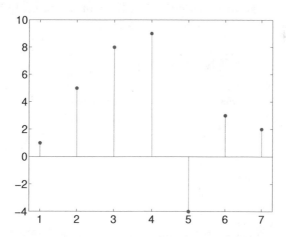

FIGURE 2.11: Example of using the MATLAB `stem` function (for x = [1,5,8,9,-4,3,2];)

Box 2.5.1: The atypical indices of MATLAB

It should be noted that MATLAB array and matrix indices start at 1 (rather than 0 as is common with other programming languages such as Python, C, C++ and Java).

2.5.3   Elementary Discrete Time Signals*

Unit Impulse

The unit impulse is a discrete time signal that is one at $(n = 0)$. It is also known as unit sample or Kronecker delta function and can be formally defined as:

$$\delta[n] \equiv \begin{cases} 1 & n = 0, \\ 0 & n \neq 0. \end{cases}$$

This can be defined and visualised by MATLAB as:

```
>>n=-7:7;
>>unitI=(n==0);
>>stem(n,unitI);
```

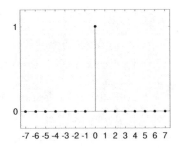

Shifted Unit Impulse

The shifted unit impulse is identical to the unit impulse, but the impulse is shifted to a new position (n_0) on the x axis

$$\delta(n - n_0) \equiv \begin{cases} 1 & n = n_0, \\ 0 & n \neq n_0. \end{cases}$$

This can be defined and visualised by MATLAB as:

```
>>n0 = 3;
>>n=-7:7;
>>sunitI=(n==n0);
>>stem(n,sunitI);
```

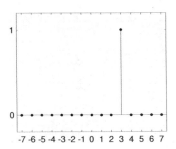

Unit Step

The Unit Step is a discrete signal that is defined as being unity at zero and above. It can be formally defined as:

$$u[n] \equiv \begin{cases} 1 & n \geq 0, \\ 0 & n < 0. \end{cases}$$

This can be defined and visualised by MATLAB as:

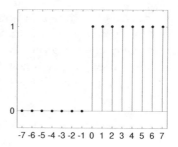

```
>>n=-7:7;
>>step=(n>=0);
>>stem(n,step);
```

Shifted Unit Step

The shifted unit impulse is identical to the unit step, but the impulse is shifted to a new position (n_0) on the x axis

$$u[n - n_0] \equiv \begin{cases} 1 & n \geq n_0, \\ 0 & n < n_0. \end{cases}$$

An example of the shifted unit step function can be defined and visualised by MATLAB as:

```
>>n0 = 3;
>>n=-7:7;
>>sstep=(n>=n0);
>>stem(n,sstep);
```

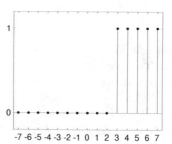

2.5.4 Periodic Signals

A periodic signal $\tilde{x}[n]$ can be defined as the periodic extension of a finite length signal $x[n]$. If the length of $x[n]$ is N and N is also taken as the period of $\tilde{x}[n]$, $\tilde{x}[n]$ can be defined as:

$$\tilde{x}[n] = x[n \bmod N], \quad n \in \mathbb{Z}, \ N \in \mathbb{N}, \ x \in \mathbb{R}, \ \tilde{x} \in \mathbb{R}, \tag{2.30}$$

where $n \bmod N$ is the modulo (remainder) operation of n given the divisor N. Periodic signals (usually denoted with a tilde, e.g., $\tilde{x}$) can be implemented within MATLAB using functions such as sawtooth, square, cos, sin, etc. These functions assume a fixed period of 2π.

Sawtooth Wave

A sawtooth wave is a periodic wave that can be defined in many different ways and is a fundamental building block for musical synthesis techniques in both its continuous and discrete forms. As both sawtooth, triangle and square waves contain a very large quantity of harmonic content they are ideal (and often used) oscillators for subtractive synthesis. One of the simplest definitions of a sawtooth wave can be defined and illustrated as:

$$x[n] \qquad\qquad \equiv -1 + 2n/N \qquad\qquad 0 \le n < N, \qquad (2.31)$$

$$\tilde{x}[n] \qquad\qquad = x[n \bmod N], \qquad\qquad\qquad\qquad (2.32)$$

This can be implemented within MATLAB using the sawtooth function and visualised using the following code:

```
>>N = 32;
>>fs = N/(2*pi);
>>t = -3*pi:1/fs:pi*3;
>>x = sawtooth(t,1);
>>stem(t,x,'filled');
```

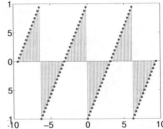

Triangle Wave

The triangle wave is also an elemental waveform (in both continuous and discrete forms) that has been used as a building block for subtractive musical synthesis. The repeating pattern of the triangle wave is a sawtooth wave back to back with an inverted sawtooth wave (the elemental forms). It can also be easily implemented using the MATLAB function sawtooth but with a window parameter of 0.5 (see the MATLAB help file document for sawtooth for a more in-depth description).

$$x[n] \qquad\qquad \equiv \begin{cases} 1 & 0 <= n < N/2, \\ -1 & N/2 <= n < N \end{cases} \qquad (2.33)$$

$$\tilde{x}[n] \qquad\qquad = x[n \bmod N] \qquad\qquad\qquad\qquad (2.34)$$

This can be defined and visualised by MATLAB as:

```
>>N = 32;
>>fs = N/(2*pi);
>>t = -3*pi:1/fs:pi*3;
>>x = sawtooth(t,0.5);
>>stem(t,x,'filled');
```

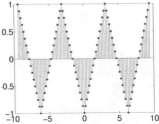

Square Wave

The square wave is also an important elemental waveform (in both continuous and discrete forms) that has been used as a building block for subtractive musical synthesis. The repeating square wave pattern can be defined in many different ways. However, in this example, the signal is either -1 or 1 with the length of each section being equal.

$$x[n] \equiv \begin{cases} 1 & 0 <= n < N/2, \\ -1 & N/2 <= n < N \end{cases} \qquad (2.35)$$

$$\tilde{x}[n] = x[n \bmod N] \qquad (2.36)$$

This can be defined and visualised by MATLAB as:

```
>>N = 32;
>>fs = N/(2*pi);
>>t = -3*pi:1/fs:pi*3;
>>x = square(t);
>>stem(t,x,'filled');
```

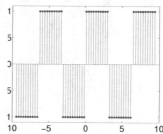

An example of musical synthesis using a square wave and the square MATLAB function is given in the last chapter of this book.

Sinusoid

A real sinusoidal sequence can be defined as the following periodic wave:

$$\tilde{x}[n] = A \cos (\omega_0 n + \phi), n \in \mathbb{Z},$$

where A is the amplitude of the wave, ω_0 is the angular frequency of the wave and ϕ is the phase offset.

This can be defined and visualised by MATLAB as:

```
>>A = 1; N = 32; phi = 0; w0 =
   1;
>>fs = N/(2*pi); % Sampling
   frequency (samples/sec)
>>t = -3*pi:1/fs:3*pi; % Time
   Vector
>>x = A*cos(w0*t+phi); % Sampled
   aperiodic triangle
>>stem(t,x,'filled');
```

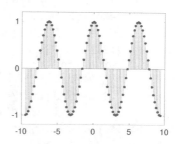

2.6 Sampling

The sampling frequency is defined as $F_s = \frac{1}{T}$ (where T_s is the sampling interval) and the sampling instances of the sequence are at $t = nT_s$ where n are the indices and t are the sampling time positions (a illustration of how a continuous signal can be sampled is shown in Figure 2.12). The continuous signal $x(\cdot)$ is therefore sampled at these temporal positions to give the discrete signal $x[\cdot]$ thus;

$$x[n] = x(n \cdot T_s) \tag{2.37}$$

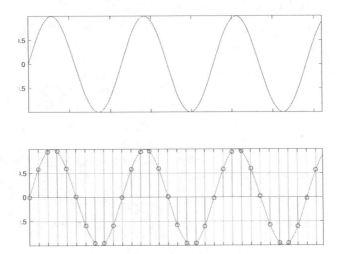

FIGURE 2.12: Sampling of a continuous/analogue signal

```
1  samplingFrequency = 44100;
2  deltat=1/samplingFrequency;
3  sinusoidFrequency = 4405;
4  tindex = 0:31;
5  tindex2 = tindex*deltat;
6  tindex3 = 0:127;
7  tindex4 = tindex3*deltat/4;
8  sampledAmp = sin(sinusoidFrequency*tindex2*2*pi);
9  Amp = sin(sinusoidFrequency*tindex4*2*pi);
10 subplot(2,1,1);
11 plot(tindex4, Amp);
```

```
12  axis tight;
13  set(gca,'xticklabel',[]);
14  subplot(2,1,2);
15  stem(tindex2, sampledAmp,'-o'); hold on;
16  plot(tindex4, Amp);
```

2.6.1 Shannon's Sampling Theorem

The sampling theorem is the key result of the seminal work on information theory by Claude Shannon [4]. The result states that an analogue signal can only be exactly reconstructed when it is sampled at a rate that is greater than half of its maximum frequency content:

$$F_{max} < \frac{F_s}{2}, \tag{2.38}$$

where F_{max} is the maximum frequency contained within the analogue signal and $F_s/2$ is half the sampling rate (F_s) also known as the Nyquist frequency. When met, this condition guarantees that the original signal can be perfectly reconstructed using carefully defined sinc interpolation functions. When this condition is not met the result is Aliasing. Aliasing is where one input frequency is confused (or aliased) with another.

2.6.2 Quantisation

Once the sampling positions in time have been defined, the resultant sampled values are still real numbers (rather than whole or rational numbers). In order to effectively transform these values to a digital representation, each sampled value will need to be represented, each with a finite number of values. Converting a real-valued continuous variable into a finite number of symbols is known as quantisation. [4]

A linear quantisation scenario can be defined (and illustrated in Figure 2.13) as follows:

- B: The number of bits to encode each sample

- R: The range of the input

- $-\frac{R}{2}$: The minimum input value

- $\frac{R}{2}$: The maximum input value

- L: The number of quantisation values = 2^B

- Q: Quantisation step size = $\frac{R}{L}$

---

[4]This is commonly done using a division and rounding operation in order that the quantised values can be represented using a limited number of binary symbols.

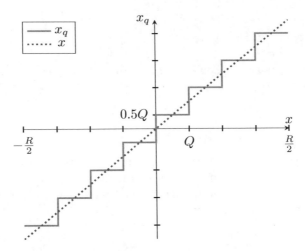

FIGURE 2.13: Example linear quantisation: x-axis is the continuous input and y-axis is the quantised output.

2.6.2.1   Quantisation Noise

The process of quantisation changes the original signal value x to its closest quantisation value x_q as illustrated by Figure 2.13. This change of value can be considered as adding "Quantisation Noise" to the signal. Denoting the quantisation error to be e_q, the quantised value x_q can be defined within the following relationship:

$$x_q = x + e_q. \tag{2.39}$$

This quantisation error (e_q) is therefore specifically dependant on the quantisation step Q and is within the range $-\frac{Q}{2} \cdots + \frac{Q}{2}$

Given the assumption (most often valid) that this noise is uniformly distributed across the range $-\frac{Q}{2} \cdots + \frac{Q}{2}$, the probability of the error $P(e)$ is given by:

$$P(e) = \begin{cases} 1/Q & \text{for} -Q/2 < e < Q/2 \\ 0 & \text{otherwise} \end{cases} \tag{2.40}$$

The signal to noise ratio (SNR) is a key analysis for quantisation noise. The signal to noise ratio is defined as the ratio of the Root Mean Squared (RMS) value of the original signal to the noise signal. If measured in decibels and it can be defined as:

$$SNR = 10\log_{10}\frac{P_x}{P_e}, \tag{2.41}$$

where P_x and P_e are the power of the signal and error, respectively. Power is defined in (2.48) or equivalently the square root of the Root Mean Squared (RMS) value.

Firstly, for a unit amplitude sine wave $P_x = 1/2$. Secondly, the power of the quantisation error is given by the following integral across the quantisation range $(-\frac{Q}{2} \cdots + \frac{Q}{2})$:

$$P_e = \int_{-Q/2}^{Q/2} P(e)e^2 de = \int_{-Q/2}^{Q/2} \frac{1}{Q} e^2 de = Q^2/12. \tag{2.42}$$

Therefore, analysing the SNR of quantisation error (given (2.41), (2.42) and $P_x = 1/2$):

$$SNR_{Q\_error} = 10\log_{10} \frac{6}{Q^2}. \tag{2.43}$$

For quantisation using a finite number of B bits 2^B numbers can be represented. To represent a unit amplitude signal in the range -1 to 1, the step size is equal to:

$$Q = \frac{1}{2^{B-1}} \tag{2.44}$$

Equating (2.43) and (2.44) gives:

$$SNR_{Q\_error} = 10\log_{10}\left(6 \cdot 2^{2B-2}\right). \tag{2.45}$$

Very approximately this gives the following rule of thumb:

$$\boxed{SNR_{Q\_error} \approx 6 \cdot B} \tag{2.46}$$

This rule of thumb indicates that for each additional bit in the digital representation the SNR increases by 6dB.

2.6.3   Energy*

The energy of a discrete time signal is defined as the sum of the square of the sampled signal values:

$$E_x \equiv \sum_{n=-\infty}^{\infty} |x[n]|^2. \tag{2.47}$$

2.6.4 Power*

The average power of a signal can be defined as:

$$P_x \equiv \lim_{N \to \infty} \frac{1}{2N+1} \sum_{n=-N}^{N} |x[n]|^2 . \tag{2.48}$$

2.7 The Z Transform

The Z transform is a key technique for the analysis of discrete time signals. It gives a way to effectively represent digital filters as a polynomial and therefore can easily solve linear problems through polynomial manipulations. It also has a key relationship with the Discrete Fourier Transform where complex values of Z are constrained to be on the unit circle. The most common definition of the Z transform is single sided (i.e., defined for positive indices within the summation) and is given by a weighted sum polynomial (with negative exponentials):

$$X(z) = x[0] + x[1]z^{-1} + x[2]z^{-2} + x[3]z^{-3} + \ldots \tag{2.49}$$

$$X(z) = \sum_{n=0}^{\infty} x[n]z^{-n} \tag{2.50}$$

In general, the values of z are complex and are often visualised on the complex Cartesian plane (the Argand diagram shown in Figure 2.4). By convention (and within the following examples) a time domain signal is represented as a lower case letter and its Z transform is represented by the same letter in upper case, i.e., if a time domain signal is x then its Z transform is X.

2.7.1 Z Transform Convolution Property

One of the main reasons that the Z transform finds so much utility is its so-called convolution property. This property defines the following relationship:

$$\text{if} \quad x_2[n] = x_0[n] * x_1[n]$$
$$\text{then} \quad X_2(z) = X_0(z) \cdot X_1(z)$$

Where X_0, X_1 and X_2 are the Z transforms of the discrete time signals x_0, x_1 and x_2, [5] i.e., the convolution of two time domain signals is equivalent to the multiplication of their Z transforms. As the Z transform is intimately related to the Fourier transform, the Z transform convolution property is directly related to the convolution theorem of the Fourier Transform.

---

[5] $\cdot$ indicates the product and $*$ indicates convolution.

Z transform: Convolution Property Illustration*

The following gives an illustrative example of the Z transform convolution property. Given two example sampled signals x_0 and x_1 their convolution can be represented as follows (firstly in the time domain and then secondly in the Z transform domain):

$$x[n] = \sum_{n=-\infty}^{\infty} x_0[k]x_1[n-k] \quad \xrightarrow{\text{Z transform}} \quad X_0(z)X_1(z).$$

The following illustrates the convolution property of the Z transform using the following example signal definitions:

$$x_0[0] = 1, \ x_0[1] = -2, \ x_0[2] = 3.$$
$$x_1[0] = 4, \ x_1[1] = -1, \ x_1[2] = 2.$$

The Z transform of x_0 is:

$$X_0(z) = 1 - 2z^{-1} + 3z^{-2}.$$

The Z transform of x_1 is:

$$X_1(z) = 4 - z^{-1} + 2z^{-2}.$$

Multiply the two Z transforms (polynomial multiplication):

$$X(z) = X_0(z)X_1(z),$$

$$X(z) = 4 - 9z^{-1} + 16z^{-2} - 7z^{-3} + 6z^{-4}.$$

We then take the inverse Z transform to give the output sampled signal:

$$x[0] = 4, \ x[1] = -9, \ x[2] = 16, \ x[3] = -7, \ x[4] = 6.$$

To see the equivalence between Z transform polynomial multiplication the same values are calculated using the standard convolution formula:

$$x[n] = \sum_{n=-\infty}^{\infty} x_0[k]x_1[n-k].$$

$$x[0] = x_0[0]x_1[0] = \hspace{6cm} 4$$
$$x[1] = x_0[0]x_1[1] + x_0[1]x_1[0] = -1 - 8 = \hspace{2.5cm} -9$$
$$x[2] = x_0[0]x_1[2] + x_0[1]x_1[1] + x_0[2]x_1[0] = 2 + 2 + 12 = \hspace{0.5cm} 16$$
$$x[3] = x_0[1]x_1[2] + x_0[2]x_1[1]] = -4 - 3 = \hspace{2.3cm} -7$$
$$x[4] = x_0[2]x_1[2] = \hspace{6cm} 6$$

This illustrates the equivalence of the methods.

MATLAB Example of Polynomial Multiplication-Convolution*

The following two pieces of MATLAB code show the equivalence of convolution in the time and Z transform domain. The second code uses MATLAB's symbolic math toolbox. Within this code z is defined as a mathematical symbol (using syms) rather than a variable. The subsequent expressions are therefore evaluated symbolically in a way similar to Maple or Mathematica. The simplify function evaluates the symbolic expression and simplifies it to the most useful form (in this case this simplified form is just the final polynomial multiplication).

```
1  >> x0 = [1 -2 3];
2  >> x1 = [4 -1 2];
3  >> conv(x0,x1)
4  ans =
5        4 -9 16 -7 6
```

```
1  >> syms z
2  >> x0 = 1 - 2* z^(-1) + 3*z^(-2);
3  >> x1 = 4 - 1* z^(-1) + 2*z^(-2)
4  >> simplify(x0*x1)
5  ans =
6  (4*z^4 - 9*z^3 + 16*z^2 - 7*z + 6)/z^4
```

2.7.2 Further Z Transform Properties*

There are numerous standard Z transform transformations and identities. Lists of these transforms can be found in standard reference texts such as [2, 3]. However, two important transforms applicable to the rest of the book are: the shift property and time reversal.

The shift property

A digital signal can be delayed by an integer number $k \in \mathbb{N}$ of samples within the Z transform domain as follows

$$
\begin{aligned}
\text{if} \quad x[n] &= x_1[n - k] \\
\text{then} \quad X(z) &= z^{-k} X_1(z)
\end{aligned}
$$

Time reversal

A digital signal can be reversed in time within the Z transform domain as follows

$$
\begin{aligned}
\text{if} \quad x[n] &= x_1[-n] \\
\text{then} \quad X(z) &= X_1(1/z)
\end{aligned}
$$

2.7.3   Z transform: Mulitrate Signal Processing

Some other multirate manipulations of digital signals can be represented by manipulation within the Z transform domain.

Upsampling (with zero insertion)

A key part of perfect reconstruction is the process of upsampling. Depending on the order of upsampling, the output is a digital signal with a number of zeros inserted in between the original samples. A digital signal upsampled in the time domain can be represented within the Z transform domain as follows:

Zero insertion (upsampling) in the time domain

$$y_0[n] = \begin{cases} x[n/M] & n \text{ is an integer multiple of } M \\ 0 & \text{otherwise} \end{cases} \tag{2.51}$$

Zero insertion (upsampling) in the Z transform domain

$$Y_0(n) = \sum_{n=-\infty}^{\infty} h[n]z^{-nM} = X\left(z^M\right) \tag{2.52}$$

Example for $M = 2$

$$y_0[n] = \{\dots, 0, x[-2], 0, x[-1], 0, x[0], 0, x[1], 0, x[2], 0, \dots\} \tag{2.53}$$
$$Y_0(z) = X\left(z^2\right) \tag{2.54}$$

2.8   Digital Filters: The Z Transform Representation

A generic digital filter is often most easily analysed using the Z transform. Within the time domain, the output of a digital filter can be thought of as a weighted combination of input and output values:

$$\begin{aligned} y[k] = \quad & b_0 x[k] + b_1 x[k-1] + \cdots + b_n x[k-n] \\ & -a_1 y[k-1] - a_2 y[k-2] - \cdots - a_m[k-m] \end{aligned} \tag{2.55}$$

where y is the output signal, x is the signal input, k is the time index, and $b_{0\dots n}, a_{1\dots m}$ are the weights of the inputs and the previous outputs. Taking the Z transform of both sides of (2.55) gives

$$Y(z) = H(z)X(z) = \frac{b_0 + b_1 z^{-1} + \cdots + b_n z^{-n}}{1 + a_1 z^{-1} + a_2 z^{-2} + \cdots + a_m z^{-m}} X(z) \tag{2.56}$$

where:

- the weights $b_{0...n}$ and $a_{1...m}$ are the filter coefficients (it can be assumed that $a_0 = 1$)

- $H(z)$ is the transfer function of the filter

- $Y(z)$ and $X(z)$ are the Z transforms of $y[k]$ and $x[k]$ respectively

- n and m represent the filter order

    - $n = 0$: IIR (Infinite Impulse Filter: all pole) filter
    - $m = 0$: FIR (Finite Impulse Filter: all zero) filter
    - $n > 0, m > 0$: IIR (Infinite Impulse Filter: pole-zero) filter

- IIR filters are also known as autoregressive filters. FIR filters are also known as moving average filters

- All pole IIR filters are the filters used in Linear Predictive Coding for audio processing described in detail in Section 9.4.

Circuit Diagrams of FIR and IIR Filters*

Figures 2.14, 2.15 and 12.14 show three circuit diagrams of digital filters. These types of circuit diagrams are used to illustrate digital filter representations that are commonly found in reference and research literature. Firstly, Figure 2.14 shows an FIR filter where there are no poles (i.e., $m = 0$). Secondly, 2.15 shows an IIR filter where there are both poles and zeros (i.e., $n > 0, m > 0$).

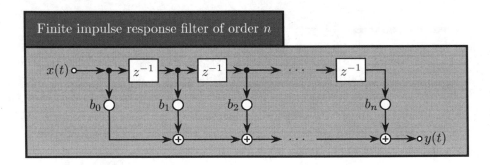

FIGURE 2.14: Finite Impulse Response (FIR): all zero digital filter.

2.8.1   Digital Filters: Poles and Zero Representation

A generic IIR filter can be represented by the transfer function (2.56). A common and more often useful representation of such a filter is illustrated below, i.e., the transfer function can be transformed into an alternative form that can

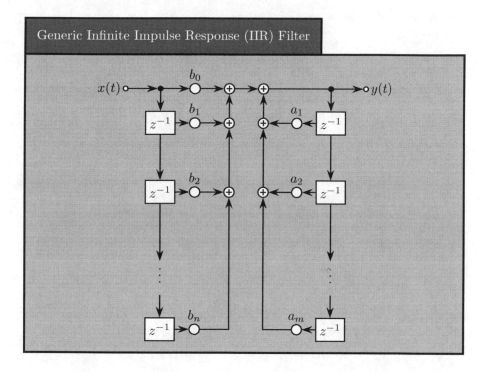

FIGURE 2.15: Infinite Impulse Response (IIR): pole-zero digital filter.

better represent the "poles" and the "zeros" of the filter. From the "fundamental theorem of algebra" both the numerator and denominator of a transfer function can be factored thus [6]:

$$H(z) = \frac{b_0 + b_1 z^{-1} + \cdots + b_n z^{-n}}{1 + a_1 z^{-1} + a_2 z^{-2} + \cdots + a_m z^{-m}},$$

$$H(z) = \frac{\left(1 - q_1 z^{-1}\right)\left(1 - q_2 z^{-1}\right)\ldots\left(1 - q_n z^{-1}\right)}{\left(1 - p_1 z^{-1}\right)\left(1 - p_2 z^{-1}\right)\ldots\left(1 - p_m z^{-1}\right)}, \qquad (2.57)$$

where q_k is the kth zero and p_k is the kth pole. Filter zeros occur when a value of z results in the top of Equation (2.57) (the numerator) being zero. Filter poles occur when a value of z results in the bottom of Equation (2.57) (the denominator) being zero. The zero and pole positions are commonly complex and when plotted on the complex z-plane it is called the pole-zero plot.

- As z approaches a pole, the denominator approaches zero and $H(z)$ approaches infinity

---

[6]Disregarding normalisation.

- As z approaches a zero, the numerator approaches zero and $H(z)$ approaches zero

- The region of convergence in the Z transform domain is the region within the unit circle

The Poles and Zeros of an Example IIR Digital Filter

An example second order (pole and zero) IIR filter is:

$$y[n] = -1/2y[n-2] + x[n] + x[n-1]. \tag{2.58}$$

Taking the Z transform of x and y on both sides of (2.58) gives:

$$Y(z) = -1/2z^{-2}Y(z) + X(z) + z^{-1}X(z). \tag{2.59}$$

Further manipulations leads to:

$$
\begin{aligned}
Y(z) + 1/2z^{-2}Y(z) &= X(z) + z^{-1}X(z), \\
Y(z)(1 + 1/2z^{-2}) &= X(z)(1 + z^{-1}), \\
Y(z) &= \frac{(1 + z^{-1})}{(1 + 1/2z^{-2})}X(z).
\end{aligned} \tag{2.60}
$$

The response of the filter of the system is defined as the transfer function $H(z)$ given that:

$$Y(z) = H(z)X(z). \tag{2.61}$$

Therefore combining (2.60) and (2.61) we can generate the following expression for the overall filter response $H(z)$.

$$
\begin{aligned}
H(z) &= \frac{1 + z^{-1}}{1 + 1/2z^{-2}} \\
H(z) &= \frac{z(z+1)}{z^2 + 1/2} \\
H(z) &= \frac{z(z+1)}{(z + i\sqrt{1/2})(z - i\sqrt{1/2})}
\end{aligned} \tag{2.62}
$$

This filter therefore has:

- Two zeros: one at $(z = 0)$ and one at $(z = -1)$

- Two poles: one at $\left(z = +\sqrt{\frac{1}{2}}i\right)$ and one at $\left(z = -\sqrt{\frac{1}{2}}i\right)$

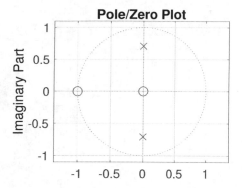

FIGURE 2.16: Pole zero plot: × is a pole, ○ is a zero.

The pole zero plot of this filter is shown in Figure 2.16. This plot is easily created using the following MATLAB code. `dfilt` is a MATLAB structure that can create a large number of digital filters. `dfilt.df2` is a function that can easily define a general IIR filter through the definition of the weights a and b. The `zplane` MATLAB function takes the filter created by `dfile.df2` and displays it on the *z* plane (amongst other functionalities).

```
1  b = [1 1];
2  a = [1 0 0.5];
3  dd = dfilt.df2(b,a)
4  zplane(dd, 'fontsize' , 14);
```

An excellent demonstration of the pole-zero view of an IIR filter is given by the zpgui MATLAB GUI developed by Tom Krauss and David Dorran illustrated in Figure 2.17. [7]

2.8.2 The Relationship Between the Z Transform and the Discrete Time Fourier Transform (DTFT)*

$$H(\omega) = \sum_{n=-\infty}^{\infty} h[n]e^{-j\omega n}$$

$$H(\omega) = \sum_{n=-\infty}^{\infty} h[n]\left(e^{j\omega}\right)^{-n}$$

The *z* transform is a generalisation of the DTFT

- i.e., substitute $z = e^{j\omega}$

---

[7] Obtained from https://dadorran.wordpress.com/2012/04/07/zpgui/.

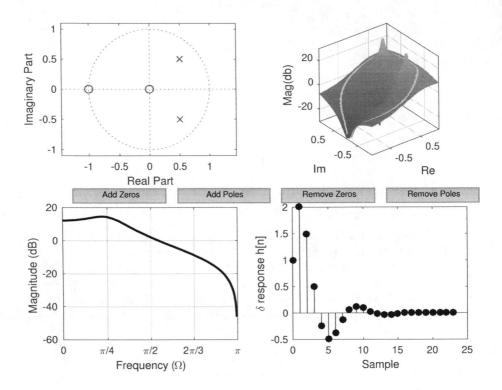

FIGURE 2.17: ZPGUI Matlab demo. [7]

$$H(\omega) = \sum_{n=-\infty}^{\infty} h[n]z^{-n}$$

2.9　Machine Learning*

In order to set the scene for understanding speech recognition systems, it is necessary to understand the science and engineering subject of recognition. This subject is most often referred to as Machine Learning.

The problem of pattern recognition is the process of discrimination or classification of an event. Humans are constantly processing sensory events, giving them an interpretation and then acting on that interpretation. Such an interpretation within a machine context is finding a structure within the input data or determining that this data is similar to some known structure.

A *pattern* in such a context can be defined as the machine input resulting from a single event. Similar structures can be grouped together to form *pattern classes*. However, similarity is a relative concept within pattern recognition. For example, the recognition of an organic life form may entail hierarchically more precise comparisons such as animal, mammal, human, etc. This is one of the main reasons why the problem on unconstrained automatic pattern recognition has proved such a difficult and intractable problem.

Conventional machine learning is comprised of two separate parts: *Feature Extraction* and *Classification*.

2.9.1 Feature Extraction

Instead of using the raw data, selected measurements called *features* extracted from the raw data are used as the basis for classification. Features should be designed to be invariant or less sensitive with respect to commonly encountered variations and distortions that are unimportant for classification, whilst containing fewer redundancies. The extraction of a set of features produces a *feature vector*, i.e., a feature vector x will take the form:

$$x = \begin{bmatrix} x_1 \\ x_2 \\ \vdots \\ x_n \end{bmatrix}. \tag{2.63}$$

Feature vectors are represented by column vectors (i.e., $n \times 1$ matrices). A feature vector can be either represented as Equation (2.63) or alternatively as its transpose $x = [x_1, x_2, \ldots, x_n]^T$.

Feature extraction is therefore the process of mapping the measurement space (M-space) of the raw data to *feature space* (X-space), where dim (X) << dim (M). The key decision is therefore which features to extract. The criterion of feature extraction can be sensibly based on either:

- the importance of the features in characterising the patterns.

- the contribution of the features to the performance of recognition.

2.9.2 Classification

Classification within machine learning involves the partition of feature space. This can be divided into two types:

Supervised learning: Supervised learning requires a separate training stage that uses training samples labelled from a set of predefined classes. Once the system has been trained by the labelled samples, it can then be used to classify new samples of unknown classification. Automatic speech recognition is a supervised learning classifier.

Unsupervised learning: In problems where the classes are not specified *a priori*, no labelled samples are available. In these situations we need to find natural groupings within the data. This is achieved through the process of *clustering* which seeks to find subsets of samples and group them into *a posteriori* classes.

2.10 Supervised Learning

Supervised learning involves separate training and classification stages.

2.10.1 Training

We define a finite number (L) of feature classes denoted by C_i, $1 \leq i \leq L$. We have a set of training samples (the *training set*) of known classification each labelled as one of the known feature classes. As we are dealing exclusively with the decision-theoretic approach the samples are feature vectors. If the training samples are feature vectors denoted by $\boldsymbol{x}^{i,j}$ then:

$$\boldsymbol{x}^{i,j} \in C_i, \quad 1 \leq i \leq L, \quad 1 \leq j \leq M_i \tag{2.64}$$

i.e., we have M_i samples of pattern class C_i. Each sample vector $\boldsymbol{x}^{i,j}$ is an element of $\mathbb{R}^n$ feature space (we assume that feature extraction has already been carried out). $\boldsymbol{x}^{i,j}$ is a feature vector having n components, denoted by $x_k^{i,j}$, $1 \leq k \leq n$.

2.10.2 Decision Rule Classification

Knowledge about the class distributions is compiled from the training set in a training stage. The classifier then uses this knowledge by transforming it into a classification rule (decision rule). The classifier is a system that takes a new sample $\boldsymbol{x}^*$ of unknown classification and assigns it one of the known pattern classes C_i, $1 \leq i \leq L$ according to the decision rule. The decision rule can be effected using either simple decision functions, correlators, optimal statistical classifiers or neural networks. An n dimensional feature space requires hypersurface decision boundaries of dimension $(n - 1)$ to separate each class. Alternatively, where there are no distinct classes to be classified, i.e., the system is designed to estimate a continuous variable, the estimation process is known as "regression".

2.11 Optimal Statistical Classification

The aim of optimal statistical classification is to produce the lowest probability of committing classification errors. In order to find a quantitative expression for the probability of committing a classification error we need an expression for the probability function that a feature vector x belongs to a class w_i. This function is the class *a posteriori* probability density function $p(w_i|x)$. This can be expressed in terms of the class likelihoods and the class *a priori* probabilities using Bayes' Formula:

$$p(w_i|x) = \frac{p(w_i)p(x|w_i)}{p(x)} \qquad (2.65)$$

where:

- $p(x|w_i)$ is the probability density function of the patterns from class w_i

- $p(w_i)$ is the (prior) probability of the occurrence of class w_i

- $p(x)$ is the *unconditional probability density* of x irrespective of class membership (also known as the evidence)

Figure 2.18 shows a two-class, one-dimensional problem where the two classes are equally likely to occur, $p(w_1) = p(w_2)$. The problem has therefore been reduced to considering the two probability density functions of the two classes. In this case the decision boundary is therefore the point x that satisfies $p(x|w_1) = p(x|w_2)$.

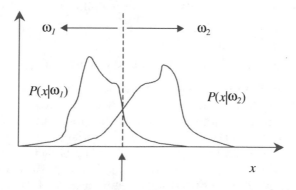

Optimal Decision Boundary

FIGURE 2.18: Probability density functions for two 1-D pattern classes. The optimal decision boundary is the value of x where the two classes are equally likely to occur (in this case).

2.11.1   Machine Learning in the Context of Speech Recognition

The classes C within a speech recognition system are the speech elements that are to be learned and recognised. These are often words, subwords (such as phones) or combinations of phones/phonemes in the form of n-grams (where n is the number of tied phones/phonemes). The feature extraction stage of a speech recognition system (in order to extract the feature vectors $x^{i,j}$) will often use perceptually invariant features such as MFCC or PLP-RASTA features. Speech recognition systems are most often implemented within a supervised classification system. In such a system many "ground truth" examples of word/phone/n-gram samples and their accurate labels are used to train the classification system. The decision as to what word or phone (or n-gram) has been said within a newly input sound is implemented using feature extraction from the candidate sound followed by a decision implemented by the classifier. Classifiers include Deep Neural Networks (e.g., Recurrent Neural Networks: RNNs), Support Vector Machines (SVMs), etc. The temporal variation of speech is often accommodated and modelled using Hidden Markov Models (HMMs). Machine learning techniques for speech recognition are described in more detail in Chapter 10.

2.12   The Analysis of Speech Signals

Speech-based audio signals have a very specific appearance within the time and frequency domains. One of the main theories that has significantly helped in the analysis of speech signals is the Source-Filter theory of speech production.

2.12.1   Source-Filter Theory of Speech Production

The source-filter theory of speech production is fundamental to the understanding of speech signals and provides the theoretical basis for the majority of current speech coding and automatic speech recognition systems.

Its origins can be traced to experiments undergone by Johannes Müller (1848) whereby he produced vibrations of larynges excised from human cadavers through blowing air through them. Through these experiments, he noticed that the sounds produced significantly differed from those commonly found in human speech. Speech-like sounds were only achieved when a tube (roughly the length of a human vocal tract) was placed over the larynx. The source-filter theory of speech production was refined into its modern accepted version by Fant [1]. An excellent summary of this theory is given by Fant himself [1].

The speech wave is the response of the vocal tract to one or more excitation signals.

The Source

The source is the basic vibration that is acted upon by the filter. For voiced sounds the source is caused by the vibration of the glottis (vocal chords/folds) producing a complex periodic wave (pressure fluctuation). The spectrum of a voiced source contains energy at the fundamental frequency of the vocal chords' vibration together with its associated harmonics (i.e., with frequencies of rational fractions and whole number multiples of the fundamental).

A voiced source can be emulated through the production of specifically shaped vibrations (in the time domain) vibrating with a defined (or analysed) fundamental frequency.

Unvoiced source vibrations are caused by the vibration of the air passing through constricted parts of the vocal tract (the outputs are known as fricatives). These source sounds are characterised by having frequency content across all the frequency range.

An unvoiced source can be emulated through the production of white (or other types of noise) across the entire frequency spectrum.

The source (as noted by Fant) can be a combination of these two types of vibrations.

The Filter

The filter is the effect of the time-varying vocal tract on the source. The filter can be considered as a combination of the effects of a small number of resonances. These resonances are called formants and can be seen as the peaks of the overall spectral envelope of the output sound. The vocal tract can therefore be considered as forming a complex time-varying filter where the formants define the peaks in the filter's frequency representation.

The filter is analysed using frequency analysis tools such as the Fast Fourier Transform (FFT) and Linear Predictive Coding (LPC) techniques. The filter can be emulated using any time-varying filter (digital or analogue) but is most commonly emulated in modern systems by LPC methods.

2.12.2   The Source-Filter Theory

The output sound-wave can be considered to be the convolution of the source and the filter, i.e., the output sound is produced by the voiced or unvoiced source signal convolved with the effect of the time-varying vocal tract "filter". This is illustrated in Figure 2.19.

It has been found that it is possible to resynthesise recognisable speech from just a model of its spectral envelope (filter). It is therefore concluded that the spectral envelope of a speech signal is more important than the source and carries the phonetic and semantic information.

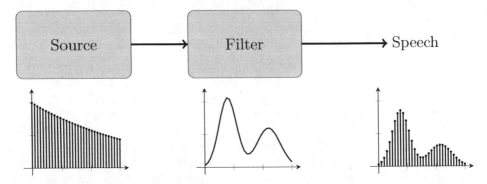

FIGURE 2.19: Illustration of the source-filter theory of speech production (time domain: top, frequency domain: bottom).

2.13 Summary

- Sinusoids are the building blocks of audio signal analysis
  - Can be derived through the formalisation of Simple Harmonic Motion (SHM)
  - SHM equation derived through the equating Newton's second law with Hooke's Law
- The Z transform is fundamental to the analysis of discrete signals
- Quantisation and sampling are the fundamental operations for the conversion from analogue to discrete signals
- The subject of machine learning is of key importance to ASR
- The source-filter theory of speech production is key for the majority of the methods discussed in the remainder of this book

2.14   Exercises

Exercise 2.1

Write some MATLAB CODE to convolve the following sequences [1 -2 3] and [4 2 -5].

Exercise 2.2

Repeat the first question, but use polynomial multiplication.

Exercise 2.3

Write some MATLAB CODE to convolve the following sequences [1 -1 2] and [3 1 -4].

Exercise 2.4

Repeat the third question, but use polynomial multiplication.

Bibliography

[1] G. Fant. *Acoustic Theory of Speech Production: With Calculations Based on X-ray Studies of Russian Articulations*, volume 2. Walter de Gruyter, 1971.

[2] A. Oppenheim and R. Schafer. *Digital Signal Processing*. Prentice-Hall, Englewood Cliffs, NJ, 1999.

[3] J. Proakis and D.K. Manolakis. *Digital Signal Processing*. Prentice-Hall, 2006.

[4] C. E. Shannon. A mathematical theory of communication. *The Bell System Technical Journal*, 27(3):379–423, July 1948.

3

Frequency Analysis for Audio

CONTENTS

> Fourier's theorem is not only one
> of the most beautiful results of
> modern analysis, but it may be
> said to furnish an indispensable
> instrument in the treatment of
> nearly every recondite question in
> modern physics.
>
> Lord Kelvin

55

3.1 Introduction

Frequency analysis is a key aspect of any speech and audio processing or manipulation technique. This chapter introduces the concepts of Fourier analysis
and its direct application to the analysis of audio signals.

3.2 Fourier Analysis

Joseph Fourier created one of the most fundamental breakthroughs in mathematics in 1807 when he asserted that an arbitrary periodic function can be
decomposed into a weighted sum of sinusoids. This is illustrated in Figure 3.1
where a square wave is approximated by a weighted sum of sinusoids. As the
number of sinusoids increases, the closer the approximation to a square wave
the resulting signal becomes. In the limit, a perfect square wave is achieved.
This figure is created by the MATLAB code:

```
1  t = linspace(-2*pi,2*pi,1000);
2  y(1,:) = sin(t);
3  y(2,:) = sin(t) + (1/3)*sin(3*t);
4  y(3,:) = sin(t) + (1/3)*sin(3*t) + (1/5)*sin(5*t) ;
5  y(4,:) = sin(t) + (1/3)*sin(3*t) + (1/5)*sin(5*t) + (1/7)*sin(7*
       t) ;
6  y(5,:) = sin(t) + (1/3)*sin(3*t) + (1/5)*sin(5*t) + (1/7)*sin(7*
       t) + (1/9)*sin(9*t);
7  plot(t,y);
```

For audio signal processing, Fourier decompositions are a key tool in
analysing audio signals as they give precise measurements of the frequency
content. This in turn is a key tool for specific audio applications (such as
speech recognition, musical synthesis, etc.). Additionally, frequency domain
processing is enabled through the use of inverse Fourier transforms; transforms that are able to take a frequency representation and invert them to
reform a time domain audio signal.

There are four specifically defined types of Fourier transform depending
on whether the signal is of a finite or infinite extent and whether it is either discrete or continuous. These four types of transform are illustrated in
Figures 3.2 and 3.3. Although the other types of Fourier transform do give
valuable insights and results for audio processing, due to the focus of this
book being discrete audio manipulation using computer-based processing,
the focus will be on the Discrete Fourier Transform (DFT) and its efficient
implementation, the Fast Fourier Transform (FFT). However, the other three

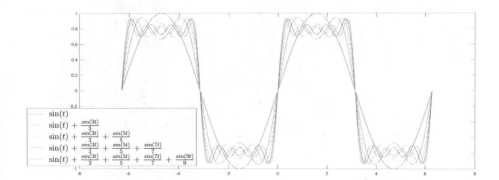

FIGURE 3.1: How a square wave can be approximated by a weighted sum of sinusoids.

Fourier transforms will now be reviewed: the Discrete Time Fourier Transform (DTFT), the Continuous-Time Fourier Transform (CTFT) and the Fourier Series (FS). For those wanting to jump to the most relevant transform for discrete audio processing, key descriptions of the Discrete Fourier Transform (DFT) and its use starts in Section 3.6 and the next sections can be skipped.

3.2.1   The Structure of Fourier Transforms and Fourier Analysis

A time domain signal is transformed into the frequency domain using a forward Fourier transform (selected from the four forms). A signal within the frequency domain can also be transformed back into the time domain using the corresponding inverse Fourier transform. Within the forms shown below the time domain signal is denoted as $x(t)$ (for a continuous signal) or $x[t]$ for a discrete signal (with t being the real or discrete-valued time index of x, respectively).

3.3   The Fourier Transform*

The Continuous-Time Fourier Transform (CTFT: usually denoted just as the Fourier Transform) is continuous and of infinite extent/non-periodic within both the time and frequency domains as illustrated in Figures 3.2 and 3.3. It is the general and most commonly used mathematical form of the four as it relates real-valued functions within a continuous space to their frequency representations. It plays a pivotal role within the functional manipulations

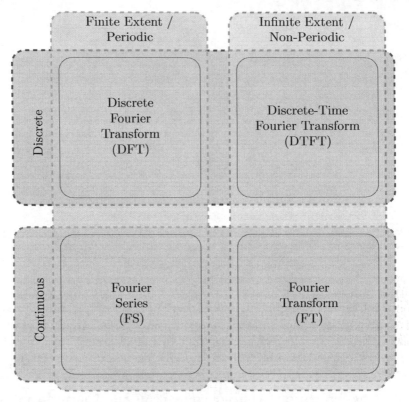

FIGURE 3.2: Time domain properties of the four types of Fourier transform.

of physics, chemistry and the statistical disciplines. Within the field of audio engineering, it is also useful in the analysis of real-world audio signals.

An arbitrary/non-periodic continuous signal $x(t)$ can be formed from the weighted integral of complex sinusoids (represented by a complex exponential). The weighting function X is itself arbitrary/non-periodic and continuous. The following expression for $x(t)$ is known as the inverse Fourier transform:

$$x(t) = \frac{1}{2\pi} \int_{-\infty}^{\infty} X(\omega)e^{j\omega t} d\omega, \tag{3.1}$$

where the radian frequency variable ω is defined as $\omega = 2\pi F$ (where F is the phasor frequency).

$X(\omega)$ is known as the (Continuous Time) Fourier Transform of $x(t)$ and can be obtained through the use of the CTFT:

$$X(\omega) = \int_{-\infty}^{\infty} x(t)e^{-j\omega t} dt, \tag{3.2}$$

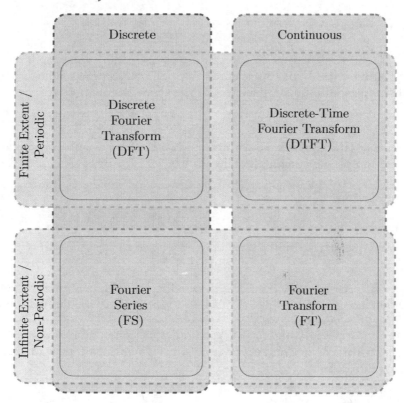

FIGURE 3.3: Frequency domain properties of the four types of Fourier transform.

x and X in this form denote a CTFT pair.

3.3.1 Parseval's Relation

A useful relationship between the CTFT pair x and X is known as Parseval's relation. This states that the total energy of the original signal x is equal to the total energy of the transformed signal X (for aperiodic signals with finite energy). For the CTFT, Parseval's relationship can be defined as:

$$\int_{-\infty}^{\infty} |x(t)|^2 \, dt = \int_{-\infty}^{\infty} |X(F)|^2 \, dF. \tag{3.3}$$

3.4   The Fourier Series*

The Fourier series transform is applicable to a periodic (i.e., defined within a finite range) continuous signal:

$$x(t) : 0 \leq t \leq T. \tag{3.4}$$

Such a periodic signal can be synthesised from a linear sum of harmonically related complex exponentials of the form $e^{jk\omega_0 t}, k = 0, \pm 1, \pm 2, \ldots$ as follows:

$$x(t) \quad = \quad \sum_{k=-\infty}^{\infty} X\left[k\right] e^{jk\omega_0 t}, \tag{3.5}$$

$$\omega_0 \quad = \quad \frac{2\pi}{T}, \tag{3.6}$$

where $X\left[k\right]$ denotes the Fourier series coefficients. As the signal x is periodic and defined in the range given by (3.4) so too are each of the summation terms in (3.5) (with period $T = 2\pi/\omega_0$).

   In order to obtain the inverse relation (and to find an expression for the Fourier series coefficients $X\left[k\right]$), some simple manipulations lead to the following integral:

$$X[k] = \frac{1}{T} \int_T x\left(t\right) e^{-jk\omega_0 t} dt, \tag{3.7}$$

$$\omega_0 = \frac{2\pi}{T}, \tag{3.8}$$

where $X\left[k\right]$ and $x(t)$ form a Continuous-Time Fourier series pair.

   It follows that the transformed signal within the frequency domain is of infinite extent but discrete:

$$X[k] : k = 0, \pm 1, \pm 2, \ldots \tag{3.9}$$

3.4.1   Parseval's Relation (for Fourier Series)

Parseval's relation (i.e., the equivalence of energy in the time and frequency domain) also holds for the Fourier series transform in this form:

$$\frac{1}{T} \int_T |x\left(t\right)|^2 dt = \sum_{k=-\infty}^{\infty} |X\left[k\right]|^2. \tag{3.10}$$

3.5   The Discrete-Time Fourier Transform*

The discrete-time Fourier transform is for the transformation of continuous-time aperiodic finite energy signals. A discrete-time forward Fourier transform of a finite energy signal can be defined as:

$$X\left(\omega\right) = \sum_{n=-\infty}^{\infty} x\left[n\right] e^{-j\omega n}. \tag{3.11}$$

In this format $X(\omega)$ represents the frequency content of the original discrete signal $x[n]$. The two differences between the Fourier transform (3.2) and the discrete-time Fourier transform are:

- The Fourier transformed signal (using (3.2)) has an infinite frequency range: $(-\infty, \infty)$ whereas the discrete-time Fourier transform has a frequency range of 2π: $(-\pi, \pi)$. ($X(\omega)$ in (3.11) is periodic with a period of 2π).

- Since the signal is discrete in nature, the discrete-time Fourier transform involves a summation rather than an integral.

Given the second point, the inverse discrete-time Fourier transform is still an integral (as with the Fourier transform), however, the integral is only over the range of $(-\pi, \pi)$.

$$x\left[n\right] = \frac{1}{2\pi} \int_{-\pi}^{\pi} X\left(\omega\right) e^{j\omega n} d\omega, \tag{3.12}$$

$$\left(n = 0, \pm 1, \pm 2, \ldots\right). \tag{3.13}$$

3.6   The Discrete Fourier Transform (DFT)

Although the three previously described transforms (the continuous-time Fourier-transform, the Fourier series transform and the discrete-time Fourier transform) are useful for the analysis of aperiodic and continuous signals (as per each definition), the main focus of this book will be the discrete Fourier transform. This is justified as:

- All of the signal processing we will consider will be discrete.

- All of the signals we will consider will be of a finite extent.

- The analysis of the frequency content of any considered discrete signal will most often need to be discrete itself for storage and further analysis within the considered discrete processing systems.

The DFT is applicable to a periodic (i.e., defined within a finite range; in this case of length N) discrete time domain signal:

$$x[n] : n = 0, 1, 2, \ldots N - 1. \tag{3.14}$$

From the definition of the DFT it follows that the transformed signal within the frequency domain is of finite extent but discrete (and of the same length as the original signal):

$$X[k] : k = 0, 1, 2, \ldots N - 1. \tag{3.15}$$

The DFT, therefore, decomposes a time domain signal (defined over a finite range) into a finite sum of weighted complex sinusoids.

$$x[n] = \frac{1}{N} \sum_{k=0}^{N-1} X[k] e^{jk\omega_0 n}, \tag{3.16}$$

$$\omega_0 = \frac{2\pi}{N}. \tag{3.17}$$

The forward discrete Fourier transform is similarly defined to the inverse transform, i.e., a finite sum of weighted complex sinusoids with the weights being the signal domain samples $x[n]$:

$$X[k] = \sum_{n=0}^{N-1} x[n] e^{-jk\omega_0 n}, \tag{3.18}$$

$$\omega_0 = \frac{2\pi}{N}. \tag{3.19}$$

3.7 Forward and Backwards Transform for the Four Types of Fourier Transforms[*]

The following table shows the four variations of the Fourier transform with their forward and backward transforms. Additionally, this table also shows the Parseval Theorem for each transform. The normalisation factor for each transform is defined variably according to different references. The normalisation here is the most commonly found, consistent and the one used in MATLAB.

Four Types of Fourier Transform

Transform	Signal, x	Spectrum, X
Fourier Transform (FT), a.k.a Continuous Time Fourier Transform (CTFT)	**continuous, aperiodic** $x(t) = \frac{1}{2\pi}\int_{-\infty}^{\infty} X(\omega) e^{j\omega t} d\omega$	**continuous, aperiodic** $X(\omega) = \int_{-\infty}^{\infty} x(t) e^{-j\omega t} dt$
Fourier Series (FS)	**continuous, periodic** $x(t) = \sum_{k=-\infty}^{\infty} X[k] e^{jk\omega_0 t}$ $\omega_0 = \frac{2\pi}{T}$	**discrete, aperiodic** $X[k] = \frac{1}{T}\int_T x(t) e^{-jk\omega_0 t} dt$ $\omega_0 = \frac{2\pi}{T}$
Discrete-Time Fourier Transform (DTFT)	**discrete, aperiodic** $x[n] = \frac{1}{2\pi}\int_{-\pi}^{\pi} X(\omega) e^{j\omega n} d\omega$	**continuous, periodic** $X(\omega) = \sum_{n=-\infty}^{\infty} x[n] e^{-j\omega n}$
Discrete Fourier Transforms (DFT)	**discrete, periodic** $x[n] = \frac{1}{N}\sum_{k=0}^{N-1} X[k] e^{jk\omega_0 n}$ $\omega_0 = \frac{2\pi}{N}$	**discrete, periodic** $X[k] = \sum_{n=0}^{N-1} x[n] e^{-jk\omega_0 n}$ $\omega_0 = \frac{2\pi}{N}$

Parseval's Theorem for the Four Types of Fourier Transform

Transform	Parseval's Theorem
Fourier Transform (FT), a.k.a Continuous Time Fourier Transform (CTFT)	$\int_{-\infty}^{\infty} |x(t)|^2 dt = \int_{-\infty}^{\infty} |X(F)|^2 dF$
Fourier Series (FS)	$\frac{1}{T}\int_T |x(t)|^2 dt = \sum_{k=-\infty}^{\infty} |X[k]|^2$
Discrete-Time Fourier Transform (DTFT)	$\sum_{n=-\infty}^{\infty} |x[n]|^2 = \frac{1}{2\pi}\int_{-\pi}^{\pi} |X(\omega)|^2 d\omega$
Discrete Fourier Transforms (DFT)	$\sum_{m=0}^{N-1} |x[n]|^2 = \frac{1}{N}\sum_{k=0}^{N-1} |X[k]|^2$

3.8 Example Discrete Fourier Transform*

The DFTs used in audio processing are often of a power of two in the order of a few hundred samples (e.g., 512 or 1024). However, the following example just uses an 8-point DFT in order to illustrate the effect of the forward transform. Firstly, the forward and backward DFT (as defined in (3.16) and (3.18)) can be formulated in matrix form:

$$X = Wx, \tag{3.20}$$

$$x = W^*X, \tag{3.21}$$

where X and x are the transform and signal vectors respectively of equal length N and W is an $N \times N$ transform matrix that performs the forward DFT. An 8-point forward DFT is defined as:

$$X[k] = \sum_{n=0}^{7} x[n] e^{-jk\omega_0 n}, \tag{3.22}$$

where $\omega_0 = \frac{2\pi}{N} = \frac{\pi}{4}$. This 8-point DFT can be thought of as a sum of 8 complex exponentials for each of 8 discrete frequencies. These 8 frequencies (indexed by k in (3.22)) can be visualised by dividing the unit circle in the complex Z-domain into 8. In general, these divisions are known as the nth roots of unity and specifically in this case the 8th roots of unity (Figure 3.4). The DFT forward transform (3.20) in this case can be illustrated using an 8×8 matrix W where each row is an analysis of a fractional frequency and each column corresponds to each point multiple within each fractional frequency decomposition:

$$
\begin{bmatrix} X[0] \\ X[1] \\ X[2] \\ X[3] \\ X[4] \\ X[5] \\ X[6] \\ X[7] \end{bmatrix}
=
\begin{bmatrix}
1 & 1 & 1 & 1 & 1 & 1 & 1 & 1 \\
1 & \sqrt{2} - i\sqrt{2} & -i & -\sqrt{2} - i\sqrt{2} & -1 & -\sqrt{2} + i\sqrt{2} & i & \sqrt{2} + i\sqrt{2} \\
1 & -i & -1 & i & 1 & -i & -1 & i \\
1 & -\sqrt{2} - i\sqrt{2} & i & \sqrt{2} - i\sqrt{2} & -1 & \sqrt{2} + i\sqrt{2} & -i & -\sqrt{2} + i\sqrt{2} \\
1 & -1 & 1 & -1 & 1 & -1 & 1 & -1 \\
1 & -\sqrt{2} + i\sqrt{2} & -i & \sqrt{2} + i\sqrt{2} & -1 & \sqrt{2} - i\sqrt{2} & i & -\sqrt{2} - i\sqrt{2} \\
1 & i & -1 & -i & 1 & i & -1 & -i \\
1 & \sqrt{2} + i\sqrt{2} & i & -\sqrt{2} + i\sqrt{2} & -1 & -\sqrt{2} - i\sqrt{2} & -i & \sqrt{2} - i\sqrt{2}
\end{bmatrix}
\begin{bmatrix} x[0] \\ x[1] \\ x[2] \\ x[3] \\ x[4] \\ x[5] \\ x[6] \\ x[7] \end{bmatrix}
\tag{3.23}
$$

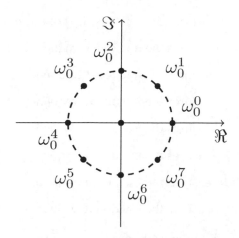

FIGURE 3.4: 8th roots of unity.

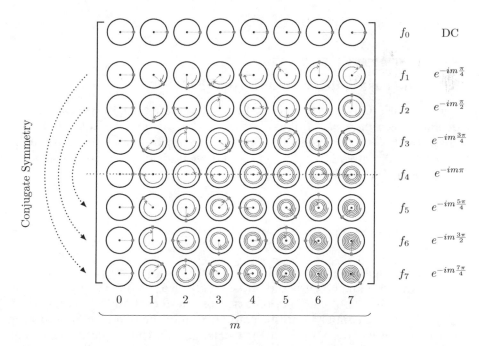

FIGURE 3.5: DFT matrix illustration: Visualisation of matrix $\mathbf{W}$ in Equation (3.22). Each row is a different frequency as indicated by f_k on the right.

To summarise the effect of the rows of **W** in the above table and Figure 3.5,

- f_0 (row 0 of **W**) measures the DC (average) of the signal
- f_1 (row 1 of **W**) measures the fractional frequency of $\frac{1}{8}$
- f_2 (row 2 of **W**) measures the fractional frequency of $\frac{1}{4}$
- f_3 (row 3 of **W**) measures the fractional frequency of $\frac{3}{8}$
- f_4 (row 4 of **W**) measures the fractional frequency of $\frac{1}{2}$
- f_5 (row 5 of **W**) measures the fractional frequency of $\frac{5}{8}$
- f_6 (row 6 of **W**) measures the fractional frequency of $\frac{3}{4}$
- f_7 (row 7 of **W**) measures the fractional frequency of $\frac{7}{8}$

Figure 3.5 shows the rotational effect of the "twiddle factor" (as shown on the right of the figure). As can be seen in this figure, due to complex conjugate symmetry, the output of $f_0 \ldots f_7$ is redundant by an approximate factor of two, e.g., the frequency represented by f_7 is the same as f_1 but is its conjugate mirror.

3.9 The Fast Fourier Transform: FFT*

The simplest implementation of a speed optimised DFT is the radix-2 decimation-in-time FFT, originally defined by Cooley and Tukey [1]. This algorithm takes a DFT of length N and recursively divides it into two equally sized (initially of length $N/2$) interleaved DFTs. The single outer stage of a length 8 FFT is illustrated in Figure 3.6.

$$X[k] = \sum_{n=0}^{N-1} x[n]\, e^{-j\frac{2\pi}{N}kn} \quad \text{where n ranges from 0 to } N-1 \tag{3.24}$$

At the first stage of the Radix-2 FFT, the input values $x[0], x[1], x[2], \ldots x[N-1]$ are separated into their even and odd elements, i.e., $(x(0), x(2), x(4), \ldots)$ and $(x[1], x[3], x[5], \ldots)$. A DFT is processed on these odd and even signals (of length $(N/2)$) separately with the combination of the two results producing the final output of the entire sequence (of length N). This single stage can then be repeated recursively leading to an overall runtime number of operations of $O(N\log(N))$.

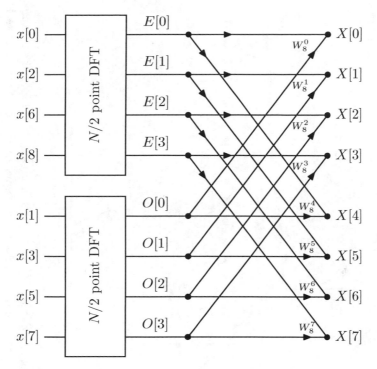

FIGURE 3.6: Circuit diagram of a one stage length 8 Radix 2 FFT (using the Cooley and Tukey [1] method).

.

A single stage of the Radix-2 algorithm can be formally represented as follows (i.e., separating the original DFT (3.24) into even and odd components).

$$X[k] = \sum_{m=0}^{N/2-1} x[2m] e^{-j\frac{2\pi}{N}k(2m)} + \sum_{m=0}^{N/2-1} x[2m+1] e^{-j\frac{2\pi}{N}k(2m+1)} \quad (3.25)$$

The reason for the speed optimisation of this formalism is the common factor multiplier.

Due to the nature of the separation of the signal into equal sized even and odd signals, the input is restricted to have a length N being a power of 2. However, simple algorithmic implementations can be used to overcome this restriction in usable implementations. The MATLAB function FFT can input signals of any length.

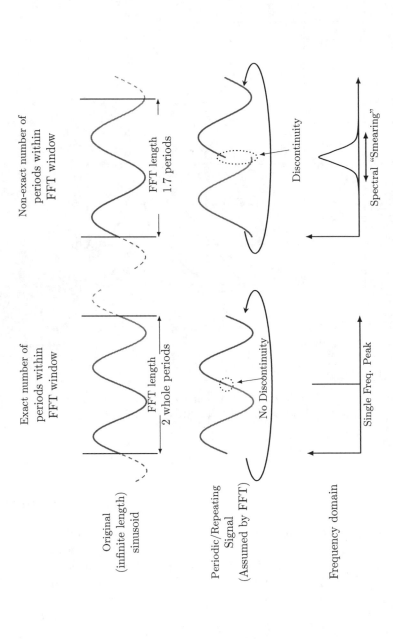

FIGURE 3.7: An illustration of spectral smearing: The left image shows an FFT length that is an exact multiple of sinusoid periods (resulting in a single frequency peak in the frequency domain) and the right image shows an FFT length of an inexact multiple of sinusoid periods (resulting in a discontinuity in the time domain and a resulting "smearing" of the frequency analysis response).

3.10 Windowing to Decrease Spectral Leakage and Improve Frequency Resolution*

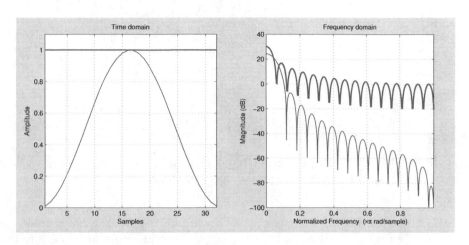

FIGURE 3.8: The square and Hanning windows to be used for FFT analysis. (Generated using the `wvtool` function.)

The DFT (and hence the FFT) uses the assumption that the signal is periodic and repeats from negative to positive infinity in time (i.e., it exists and repeats for all time). For any practical analysis of digital frequencies, a short section or segment of a digital signal must be analysed. If an FFT transform of this short signal section is transformed, the transform (erroneously) assumes that it repeats infinitely before and after the analysed segment. This erroneous assumption leads to edge effects between repeating segments and therefore to what is known as spectral leakage (a lack of frequency resolution caused by spectral information "leaking" from one frequency position into adjacent values). However, there is no way to completely avoid this "leakage" effect but through the careful use of windowing functions the effect can be significantly reduced. An illustration of the effect of using a finite FFT can be seen in Figure 3.7.

This figure shows that when selecting a limited section (i.e., truncation) of an infinitely long sinusoid, a discontinuity will appear between the repeating ends resulting in a "jump" between repeats. This "jump" will result in spurious frequency content.

This effect can also be understood through considering the segmentation (or truncation) of the original signal to a small section to be just the multiplication of the original signal with a square truncation signal.

A simple illustration of how the use of a typical windowing function

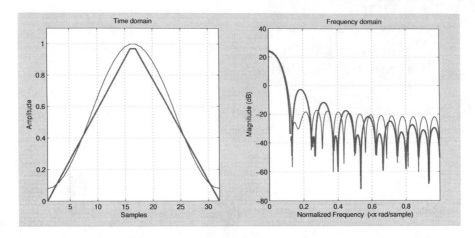

FIGURE 3.9: The triangle (Bartlett) and Hamming windows to be used for FFT analysis. (Generated using the `wvtool` function.)

(the Hanning window) is shown in Figure 3.10. [1] The original signal is a small segment (256 samples) from an inbuilt MATLAB audio file (found in the .mat file gong.mat). The bottom right subfigure shows how the windowing function reduces spectral smearing. In the case (as with this illustration) where the original signal contains some strong harmonic content, the frequency response default square windowing function will create significant side lobe smearing across the nearby frequencies. When using any form of windowing (this case using the Hanning window) such smearing is reduced leading to significant attenuation of non-harmonic content and improved frequency resolution. Figure 3.11 shows another example using a segment from the train.mat signal. These figures can be created using the following code:

```
1   load gong.mat
2   L = 256; nVals = 0:L-1;
3   sig = y(256+nVals+1);
4   subplot(221);
5   plot(sig);
6   title('Original Signal');
7   subplot(222);
8   window = hanning(256);
9   plot(window, 'g');
10  title('hanning(512)');
11  subplot(223)
12  plot(sig .* window, 'r');
13  title('Signal .* hanning(256)');
```

---

[1] The MATLAB `wvtool` function can be used to visualise a selection of windowing functions available. Typical outputs of `wvtool` are illustrated in Figures 3.8 and 3.9.

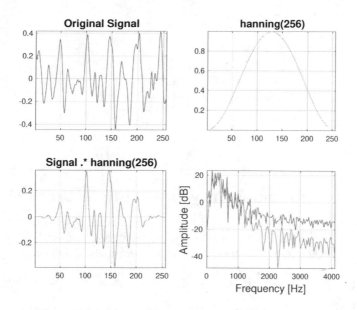

FIGURE 3.10: Illustration of how spectral smearing is reduced using a windowing function. The top left function is the original function, the top right function is the Hanning widow (of same length). The bottom left subfigure shows the dot product between the original signal and the window. The bottom right subfigure shows the frequency response of the original signal and the signal multiplied by the window.

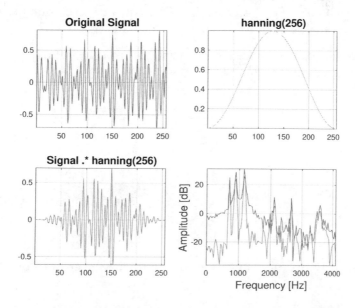

FIGURE 3.11: A similar illustration to Figure 3.10 but with different frequency content (segment from `train.mat`).

```
14  subplot(224)
15  FFT1 = fft(sig);
16  FFT2 = fft(sig .* window);
17  freq = (0:128) / 128 * Fs / 2;
18  plot(freq, 20 * log10(abs(FFT1(1:129)))); hold on;
19  plot(freq, 20 * log10(abs(FFT2(1:129))), 'r');
```

3.11   The Fast Fourier Transform (FFT): Its Use in MATLAB[*]

The discrete Fourier transform is of key importance to audio analysis applications. This section, therefore, analyses its typical use within MATLAB. The practical characteristics of the FFT can be summarised as follows:

- The FFT assumes the signal is of infinite extent (and periodic).

- The number of points in the time domain equals the number of points in the frequency domain.

- The upper half of the frequency domain output (negative frequencies) of the FFT is an aliased version of the lower half (positive frequencies). This is illustrated in Figure 3.5.

- The frequency resolution of an FFT is $F_\Delta = F_s/N$, e.g., if $F_s = 44100$ Hz and N = 1024, $F_\Delta = 43.0664$Hz. FFT frequency resolution (as applied to speech analysis) is covered in more detail in Section 9.2.2.

- The output of an FFT can be re-arranged using a shifting operation such as the MATLAB function fftshift.

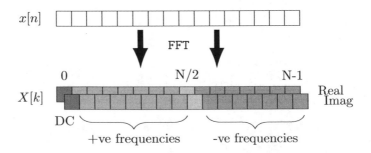

FIGURE 3.12: The use of FFT within MATLAB for the analysis of a 1D signal.

3.11.1   Matlab FFT use for 1D Signals

Figure 3.12 shows the result of an example FFT transform of a real 1D signal (of length 16 samples).

An example implementation of an FFT transform shown in this figure (with a 16-point signal) in Matlab is:

```
>> x = rand(16,1);
>> X = fft(x)

X =

   8.1213 + 0.0000i
   1.8329 - 0.5308i
  -1.4649 - 1.0749i
  -0.4158 - 0.3516i
   0.3288 + 0.6983i
  -1.5497 - 0.0254i
   0.3767 - 1.1969i
   0.4066 - 0.7716i
   0.6530 + 0.0000i
   0.4066 + 0.7716i
   0.3767 + 1.1969i
  -1.5497 + 0.0254i
   0.3288 - 0.6983i
  -0.4158 + 0.3516i
  -1.4649 + 1.0749i
   1.8329 + 0.5308i
```

This example shows that the output of an FFT transform of a real signal is complex (i.e., has real and imaginary components). It also shows the redundancy of the transform, i.e., the output above $N/2$ is a reflected conjugate copy of the signal below $N/2$. This conjugate redundancy is illustrated on the left of Figure 3.5.

It is also common to take the absolute value of the FFT output in order to obtain the magnitude of the frequency response at that point within the signal. This is also often combined with the shifting of the origin of the FFT output using the Matlab function fftshift so DC is in the middle of the output and the frequency of the output increases as the distance increases from this central output.

This implementation is shown in Figure 3.13. An actual example of such an implementation in Matlab is:

```
>> x = rand(16,1);
>> X = fft(x)
>> aX = fftshift(abs(X))
```

aX =

    0.5926
    0.5758
    0.5421
    1.1279
    1.9947
    0.9623
    1.7366
    0.5153
    6.5845
    0.5153
    1.7366
    0.9623
    1.9947
    1.1279
    0.5421
    0.5758

The FFT transform is an example of a "perfect reconstruction" transform, i.e., the input signal is exactly recovered when the transform is directly inverted. This is illustrated in Figure 3.14 where the outut $\hat{x}[n]$ equals the input $x[n]$.

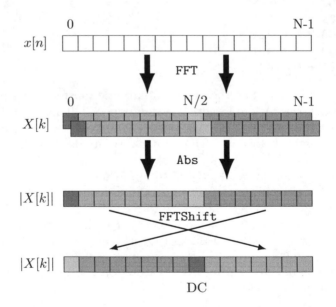

FIGURE 3.13: A typical use of the FFT transform within MATLAB to analyse the frequency of a 1D signal.

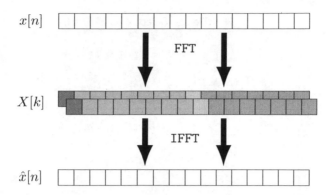

FIGURE 3.14: The forward and backward FFT transform within MATLAB.

An actual implementation example of such an implementation in MATLAB is:

```
>> x = rand(16,1);
>> X = fft(x);
>> xhat = ifft(X);
>> sum(x-xhat)
ans =
   2.2204e-16
```

3.11.2 The FFT Frequency Domain

For a signal of length N an FFT results in a complex valued signal of exactly the same length (N) as shown in Figures 3.12, 3.13 and 3.14. As indicated, these N complex outputs represent the frequency content at a specific frequency.

Plotting the Raw FFT Values

The following MATLAB code shows an example of how to visualise the FFT samples of an audio signal (with the output graph shown in Figure 3.15). Firstly, an example audio signal (the audio sample of Handel's Messiah included within MATLAB) is loaded and an initial subset is defined within the MATLAB environment. Defining the length of the signal to be N, the output axis initially is indexed from 0 to $N - 1$ (representing the N output FFT values). However, the largest index represents the same frequency as the sampling frequency F_s. In order to obtain the actual frequency of the x-axis we must divide the sampling frequency evenly across the N indices. Furthermore, in order to visualise the output of the FFT the coefficients must be converted to their magnitude (abs(X)).

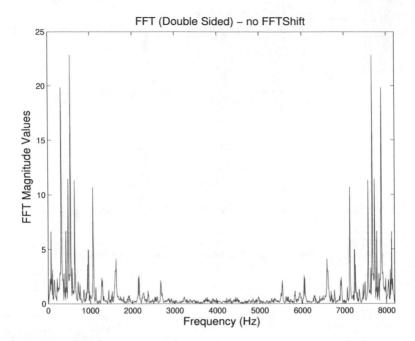

FIGURE 3.15: The magnitude of an FFT output of 1024 length subset of the inbuilt MATLAB Handel audio sample.

Figure 3.15 was created using the following code:

```
1  load handel.mat
2  L = 1024; nVals = 0:L-1; x = y(nVals+1); f = Fs*(nVals)/L;
3  X = fft(x);
4  plot(f, abs(X));
5  title('FFT (Double Sided) - no FFTShift');
6  xlabel('Frequency (Hz)');
7  ylabel('FFT Magnitude Values');
```

Plotting the Raw FFT Values Against Normalised Frequency

It is sometimes useful to normalise the frequency axis (the x-axis). In this form, we want to normalise the sampling frequency to 1 resulting in a normalised frequency in units of cycles/sample. Such a double-sided plot of normalised frequency output is shown in Figure 3.16.

> **Normlised Frequency**
>
> Normalised frequency is defined in units of cycles/sample. Therefore a plot from 0 to the sampling frequency (F_s) will be from 0 to 1 in normalised frequency. In many cases MATLAB plots normalise the frequency in units of half-cycles/sample as often only half the frequency range of from 0 to $F_s/2$ is visualised. Therefore a plot from 0 to half the sampling frequency ($F_s/2$) will be from 0 to 1 in half-cycles/sample normalised frequency. Figures such as 3.16, 3.18 and 3.19 show normalised frequency of cycles/sample and figures such as 3.20 show normalised frequency of half-cycles/sample (common in the built-in frequency plots of MATLAB).
>
> An alternative normalised frequency is radians/sample where 2π is equivalent to the sampling frequency F_s.

Figure 3.16 was created using the following MATLAB code:

```
1  load handel.mat
2  L = 1024; nVals = 0:L-1; x = y(nVals+1); f = Fs*(nVals)/L;
3  X = fft(x);
4  plot(f, abs(X));
5  title('FFT (Double Sided) - no FFTShift');
6  xlabel('Frequency (Hz)');
7  ylabel('FFT Magnitude Values');
```

A related frequency index is the use of angular frequency (in units of radians/second). This type of indexing equates the sampling frequency (F_s) to 2π. The full frequency range of the FFT output is therefore $[0, 2\pi]$ (or equivalently $[-\pi, \pi]$... see below). This type of indexing of the frequency axis will be the most commonly used within this book. An example of this type of indexing is shown in Figure 3.17 (using the same code as above).

Plotting the FFT Magnitude with `fftshift`

Due to mathematical convenience, the arrangement of the FFT output has positive and negative frequencies side by side in a slightly non-intuitive order (the frequency increases from DC until N/2 and then it decreases for negative frequencies). It is often clearer to shift the negative frequencies from the end to the start of the FFT output (making DC the centre of the output). This is achieved using the MATLAB function `fftshift` and is illustrated in Figure 3.18 and the following code:

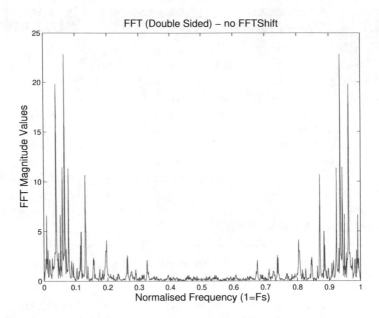

FIGURE 3.16: The magnitude of an FFT output of 1024 length subset of the inbuilt MATLAB Handel audio sample: Normalised Frequency.

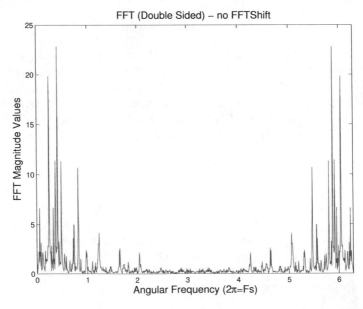

FIGURE 3.17: The magnitude of an FFT output of 1024 length subset of the inbuilt MATLAB Handel audio sample: Frequency axis normalised to angular frequency (rads/sample).

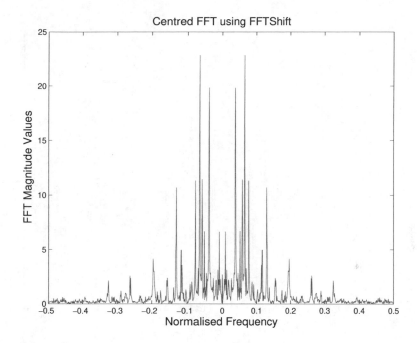

FIGURE 3.18: The magnitude of an FFT output of 1024 length subset of the inbuilt MATLAB Handel audio sample: Normalised frequency (centred using `fftshift`)

```
1  load handel.mat
2  L = 1024; nVals = 0:L-1; x = y(nVals+1); f=(-L/2:L/2-1)/L;
3  X = fft(x);
4  plot(f, fftshift(abs(X)));
5  title('Centred FFT using FFTShift');
6  xlabel('Normalised Frequency')
7  ylabel('FFT Magnitude Values');
```

Plotting the FFT Power Spectral Density (Single Sided)

As the positive and negative frequency magnitudes are identical for a real signal it is common just to plot the positive frequencies from 0 to $F_s/2$ (in normalised frequency). Additionally, the power spectral density (the square of the magnitude) is commonly displayed and this is done most often with a logarithmic y index (in dBs).

Such a single-sided plot of FFT power spectral density is shown in Figure 3.19 and within the following MATLAB code.

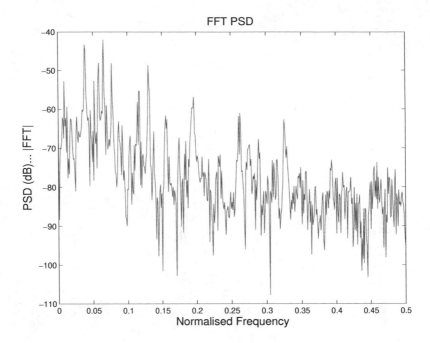

FIGURE 3.19: The Power Spectral Density of an FFT output of 1024 length subset of the inbuilt MATLAB Handel audio sample: Single-sided (positive frequencies only) in normlised frequency units.

```
1  load handel.mat
2  L = 1024; nVals = 0:L-1; x = y(nVals+1); f=(0:L/2)/L;
3  X = fft(x);
4  PSD= 1/(L*Fs)*abs(X(1:L/2+1)).^2;
5  plot(f,10*log10(PSD));
6  title('FFT PSD');
7  xlabel('Normalised Frequency');
8  ylabel('PSD (dB)... |FFT|');
```

Power Spectral Density (PSD) Estimation

The Power Spectral Density of a digital signal is defined as the square of the frequency magnitude (measured using a DFT, FFT, etc.). In many circumstances, it makes more sense to estimate and visualise the PSD of the signal rather than just its magnitude. There are a number of different methods within MATLAB to easily estimate the PSD of a signal including freqz, periodogram and pwelch. We now examine the outputs of the first two being broadly similar. The pwelch function is also a popular PSD estimator but it averages over a sliding window and is therefore slightly different than the other two methods.

Freqz
Although `freqz` is commonly used to obtain a frequency analysis of an arbitrary IIR filter it can be quickly used to visualise a one-sided frequency analysis of a real audio signal. Figure 3.20 and the following code show how this can be achieved.

```
>>load handel.mat
>>L = 1024;
>>freqz(y,1,L);
```

Periodogram
Periodogram is more commonly used to generate a quick estimate of the PSD of a signal and gives broadly similar results to the `freqz` (as they both use FFT magnitudes for PSD estimation). The periodogram function, however, is more flexible and is able to visualise a one-sided (for real signals) and two-sided (for complex signals) frequency analysis of an audio signal. Figure 3.21 and the following code show how this can be achieved.

```
>>load handel;
>>L = 1024;
>>periodogram(y,[],'onesided',L ,Fs);
```

3.12 Zero Padding the FFT

For an entire signal or within spectrograms/STFTs (see below), zero padding at the end of a signal produces a longer FFT. Such a longer FFT results in a larger number of bins and therefore coefficients that are more closely spaced in the frequency domain (i.e., the frequency resolution is higher). There is no extra information as a non-zero padded signal effectively gives a "Sinc function" interpolation within the frequency domain. However, the result may look smoother and is therefore sometimes preferred (or at least given as an option) for spectrograms/STFTs. This may be of specific use to localise the frequency of a peak in the frequency domain. However, it is in effect just a principled method of coefficient interpolation in the frequency domain.

When using the `spectrogram` function within MATLAB, if the window length is less than the FFTlen parameter, zero padding is effectively being used. Finally, zero padding can just be considered as a special windowing case, where the coefficients outside the "actual window" are set to zero.

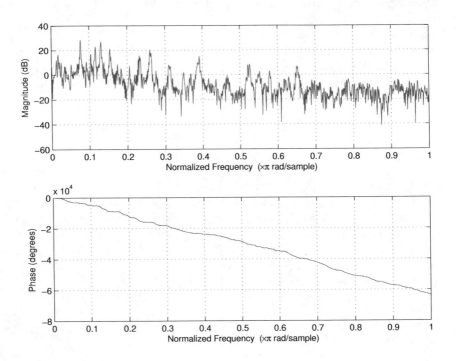

FIGURE 3.20: The FFT magnitude (and phase) of a signal using `Freqz`.

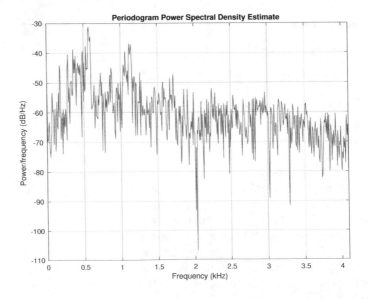

FIGURE 3.21: The FFT magnitude of a signal using `periodogram`.

3.13 Short-Term Fourier Transform: STFT

All of the Fourier transforms discussed up to this point have assumed that the spectral statistics are stationary (i.e., they do not vary in time) and therefore can be effectively measured using transforms that are periodic or infinite in length. In this section, all the previously considered transforms assume that the signal as a whole will be windowed, transformed and analysed. In order to analyse how the frequency content of signal varies with time, a spatial-frequency transform is required. The STFT is such a transform and can be defined similarly to the DFT (but with an integrated and moveable windowing function):

$$X_m(\omega) = \sum_{n=-\infty}^{\infty} x[n]w[n - mR]e^{-j\omega n} \tag{3.26}$$

where

R : The hop length (in samples)

m : The controlling parameter for shifting the window.

$X_m(\omega)$: is the STFT of the windowed data $x[n]$

$x[n]$: is the input signal indexed by n (commonly time)

$w[n]$: is the window function of length M, e.g., a Hanning window

The windowing function $w[n - mR]$ serves the same purpose as the window functions described above, but has the additionally property that it can be spatially localised and moved using the controlling spatial parameter m. The signal $f_n[m] = x[n]w[n - mR]$ is a short windowed segment of the original signal.

3.13.1 The Discrete STFT: The Practical Implementation of the STFT

As the window function is of finite extent (i.e., it is zero outside of a finite range), the STFT can be implemented using a finite length DFT. Recalling the DFT is defined as:

$$X[k] = \sum_{n=0}^{N-1} \hat{x}[n] e^{-jk\omega_0 n} \tag{3.27}$$

$$\omega_0 = \frac{2\pi}{N} \tag{3.28}$$

The discrete STFT can be defined as the forward DFT given that the input signal $\hat{x}[n]$ is actually the finite length windowed signal $x[n + mR]w[n]$ (this is equivalent to the short windowed segment $f_n[m] = x[n]w[n - mR]$ defined in (3.26)). This leads to the definition of the discrete STFT:

$$X_m[k] = \sum_{n=0}^{N-1} x[n + mR]w[n]e^{-jk\omega_0 n} \qquad (3.29)$$

$$\omega_0 = \frac{2\pi}{N} \qquad (3.30)$$

where

N : is the length of the DFT

k : is the frequency index of the output STFT: $k = 0, 1, \ldots, N - 1$

R : The hop length (in samples)

m : The controlling parameter for shifting the window

$X_m[k]$: is the STFT of input signal $x[n]$

$x[n]$: is the input signal indexed by n (commonly time)

$x[n + mR]$: is the shifted input signal indexed $n + mR$ (commonly time)

$w[n]$: is the window function of length M, e.g., a Hanning window

The forward discrete STFT in (3.29) is defined for integer values m. It is generally assumed that the hop size R in this equation is less than the length of the DFT (N). The forward discrete STFT therefore results in a grid of results indexed by m and k. This grid represents the time-frequency analysis of the signal with m representing the time axis and k representing the frequency axis. The time-frequency grid can be visualised as illustrated in Figure 3.22.

Within a practical implementation of the discrete STFT integer values of m are used that (combined with the hop size R and the length of the DFT N) span the entire signal. The structure of a STFT transform is illustrated in Figure 3.23.

3.13.2 Inverse Discrete STFT

For frequency-based processing (such as the phase vocoder) frame-based sub-band manipulation will result in potential discontinuities when recombining the inverse STFT. For this reason (and to enable perfect reconstruction discussed below) weighting windows are used in the inverse of the Discrete STFT. This system is known as the Weighted OverLap-Add (WOLA) system and can be expressed as:

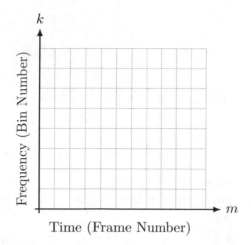

FIGURE 3.22: Time frequency tiling using the STFT. Time resolution is indicated by vertical line spacing. Frequency resolution is indicated by horizontal line spacing. The areas of each block can be rectangular in the general case and are bounded by the minimum time-bandwidth product.

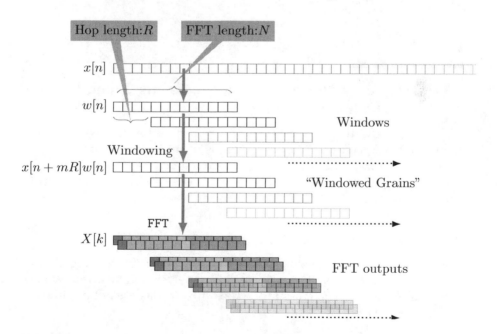

FIGURE 3.23: Time frequency processing: Illustrating hop size versus FFT length.

$$y[n] = \sum_{m=-\infty}^{\infty} y_m[n - mR]w[n - mR] \qquad (3.31)$$

$$\text{where } y_m[n] = \frac{1}{N}\sum_{k=0}^{N-1} Y_m[k]\, e^{jk\omega_0 n} \qquad (3.32)$$

$$\omega_0 = \frac{2\pi}{N} \qquad (3.33)$$

Y_m : the input to the inverse STFT. For no processing $Y_m = X_m$

y_m : Inverse DFT of Y_m

y : Inverse STFT of x

$$(3.34)$$

Although this is expressed as an infinite sum, it is effectively implemented only over the actual length of the original signal (or the length of the signal to be reconstructed).

3.14   Frequency Domain Audio Processing Using the STFT and the Inverse STFT

The forward and the inverse discrete STFT can be formulated in a way as to give perfect reconstruction when there is no intermediate processing. The most effective way to do this is to include a windowing function in the synthesis part as shown in the section above describing Weighted OverLap-Add (WOLA). WOLA is illustrated in Figure 3.24.

3.14.1   Perfect Reconstruction of the Discrete STFT

The simplest way of achieving perfect reconstruction of the Discrete STFT-Inverse Discrete STFT system is to make the analysis and synthesis windows identical. In this case (and assuming the DFT outputs are not modified) the WOLA system's output is given by:

$$y[n] = \sum_{m=-\infty}^{\infty} x[n]w^2[n - mR] \qquad (3.35)$$

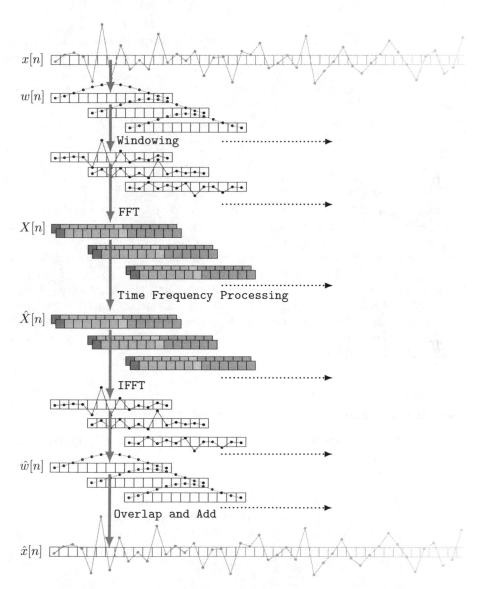

$x[n]$

$w[n]$

Windowing

$X[n]$

FFT

$\hat{X}[n]$

Time Frequency Processing

IFFT

$\hat{w}[n]$

Overlap and Add

$\hat{x}[n]$

FIGURE 3.24: Overlap and add time-frequency processing.

Perfect reconstruction can then be achieved using the condition

$$\sum_{m=-\infty}^{\infty} w^2[n - mR] = 1 \qquad (3.36)$$

This condition is sometimes referred to as the Constant OverLap-Add (COLA) condition (applied here on $w^2[n]$). It is dependent on the ratio of the frame hop R to the DFT length N. [2]

In order to ensure the COLA condition (for a system with both analysis and synthesis identical windowing filters) both the analysis and synthesis windows are weighted with $1/\sqrt{K}$, where K is defined as:

$$K = \frac{\sum_{n=0}^{N-1} w^2[n]}{R}. \qquad (3.37)$$

Figures 3.25 and 3.26 show the effect of weighting the analysis and synthesis windows to ensure that the COLA condition (and therefore perfect reconstruction) is met. These figures show three different hop sizes (relative to the DFT length) and illustrate each of the window functions $w[n-mR]$ (as defined above) as the dotted red lines. The entire squared sum $\sum_{m=-\infty}^{\infty} w^2[n - mR]$ featured in the COLA condition is shown as the solid blue line. This line can be seen to exactly equal unity for the majority of the filtering. Furthermore, these figures illustrate how, when the hop size is decreased, the windows are scaled smaller thus resulting in net combination in (3.36) of unity (and thus satisfying the COLA condition).

3.15 MATLAB Implementation of Perfect Reconstruction Discrete STFT

The following code shows MATLAB code that implements perfect reconstruction time-frequency processing using the forward and inverse Discrete STFT. This code can, therefore, be the basis of frequency based manipulation of an audio signal. In the central part of the "grain" processing loop, the magnitude and/or phase can be manipulated before the transform of the grain is inverted and the output signal is recomposed.

---

[2] Without further normalisation, common windows have a defined ratio of (R to N) to enable perfect reconstruction. For instance, the triangular (or Barlett) window has a ratio: R/N of $1/2$. Furthermore, the Hanning window has a ratio R/N of $1/4$. These choices of window and R/N ratio ensure the COLA condition and therefore perfect reconstruction. It should be noticed in these cases that the reconstruction is done without synthesis weighting and therefore the COLA condition is adjusted accordingly.

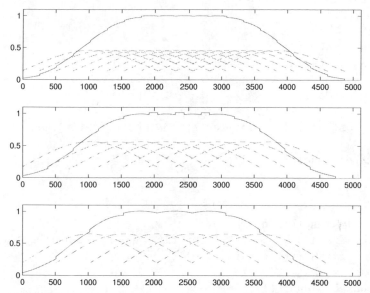

FIGURE 3.25: COLA condition satisfied by weighted Kaiser window (with varying hop size).

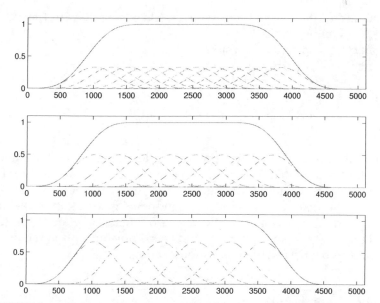

FIGURE 3.26: COLA condition satisfied by weighted Hanning window (with varying hop size).

```
1  load handel; xIn = y;
2  L = length(xIn);
3  fshift = 128;
4  lenW = 2048;
5  window1 = hann(lenW);
6  ratioCOLA =sum(window1.^2)/fshift;
7  window1 =window1/sqrt(ratioCOLA);
8  window2 = window1;
9
10 xIn = [zeros(lenW, 1); xIn; zeros(lenW-mod(L,fshift),1)];
11
12 xOut = zeros(length(xIn),1);
13 pntrIn = 0;
14 pntrOut = 0;
15
16 while pntrIn<length(xIn) - lenW
17         thisGrain = window1.*xIn(pntrIn+1:pntrIn+lenW);
18         f = fftshift(fft(thisGrain));
19         magF = abs(f);
20         phaseF = angle(f);
21         fHat = magF.* exp(i*phaseF);
22         thisGrain = window2.*real(ifft(fftshift(fHat)));
23
24         xOut(pntrOut+1:pntrOut+lenW) = thisGrain + xOut(pntrOut
              +1:pntrOut+lenW);
25         pntrOut = pntrOut + fshift;
26         pntrIn = pntrIn + fshift;
27 end
28 plot(xIn-xOut);
29 xOut = xOut(lenW+1:lenW+L) / max(abs(xOut));
30 soundsc(xOut, Fs);
```

In the central part of the "grain" processing loop, the magnitude and/or phase can be manipulated before the transform of the grain is inverted and the output signal is recomposed. This is illustrated by the next (almost identical) piece of MATLAB code whereby a "bandpass" medication to the audio is implemented. This is done by selecting a frequency region of the transformed signal and making the magnitude zero outside of this region thus implementing a bandpass frequency modification.

Listing 3.1: STFT Test Code

```
1  load handel; xIn = y;
2
3  L = length(xIn);
4  fshift = 128;
```

```
 5 lenW = 2048;
 6 window1 = hann(lenW);
 7 ratioCOLA =sum(window1.^2)/fshift;
 8 window1 =window1/sqrt(ratioCOLA);
 9 window2 = window1;
10 highpassmaxfreq = pi/2;
11 highpassmaxbinpos = round((lenW/2)*(highpassmaxfreq)/(2*pi));
12
13 xIn = [zeros(lenW, 1); xIn; zeros(lenW-mod(L,fshift),1)];
14 xOut = zeros(length(xIn),1);
15 pntrIn = 0;
16 pntrOut = 0;
17
18 while pntrIn<length(xIn) - lenW
19        thisGrain = window1.*xIn(pntrIn+1:pntrIn+lenW);
20        f = fftshift(fft(thisGrain));
21        magF = abs(f);
22        magF(lenW/2-highpassmaxbinpos:lenW/2+highpassmaxbinpos)
              =0;
23        phaseF = angle(f);
24        fHat = magF.* exp(i*phaseF);
25        thisGrain = window2.*real(ifft(fftshift(fHat)));
26
27        xOut(pntrOut+1:pntrOut+lenW) = thisGrain + xOut(pntrOut
              +1:pntrOut+lenW);
28        pntrOut = pntrOut + fshift;
29        pntrIn = pntrIn + fshift;
30 end
31 soundsc(xOut, Fs);
```

3.16 Audio Spectrograms

The analysis of the frequency content of an audio signal is usually performed on small contiguous subsets of the whole of the signal. This is because frequency content cannot be assumed to be stationary over the entire period of an audio recording (over seconds, minutes or hours). Although it is still a large assumption (especially for "plosive" type sounds) it has been found that the frequency statistics of an audio signal are better assumed to be stationary with small temporal subsets of the original. These subsets are usually known as "grains" and are of the order 10 to 30ms.

Figure 3.27 shows how a typical spectrogram is created for an audio signal. Firstly the signal is segmented into overlapping grains. The sample-wise product of these grains and a windowing function (typically a Hamming or Hanning window) then forms the windowed grain. This is then transformed into the frequency domain via a FFT/DFT, often with zero padding. A magnitude to colour mapping is then displayed of the stacked vectors generating a spectrogram image as illustrated in Figure 3.27.

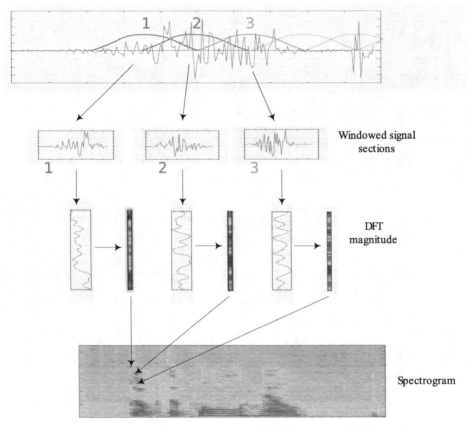

FIGURE 3.27: Creation of a spectrogram of an audio signal.

3.17 Spectrogram Using MATLAB

A Short-Time Fourier Transform (STFT) analysis of an audio signal can be easily achieved using the MATLAB function spectrogram function. [3] A spectrogram of an audio signal can be directly visualised using the function's default settings by just calling the function without any output or input parameters (other than the signal) thus [4]:

```
>> load handel.mat
>> spectrogram (y)
```

3.17.1 Common Uses of the MATLAB Spectrogram Function

Figure 3.28 shows a spectrogram generated through the use of the spectrogram function without any parameters. In most cases, the use of such default parameters is not necessarily useful. The following code illustrates a more useful use of the spectrogram function [5]:

```
1  load handel;
2  yLen = length(y);
3  windowLen = 256;
4  overlapLen = 250;
5  fftLen = 256;
6
7  spectrogram(y,hanning(windowLen),overlapLen,fftLen, Fs, 'yaxis')
```

This usage of the spectrogram function also does not have any outputs, but takes in a windowing function (of set length), an overlap number (i.e., R) and the length of the FFT itself. Such a use of the function can be defined as:

```
spectrogram(x,window,noverlap,nfft);
```

where window is the defined window (e.g., Hanning, Hamming, Kaiser, etc.), noverlap is the overlap number and nfft is the length of the FFT.

3.17.2 Aphex Twin Spectrogram Example

It is often useful to use the spectrogram function to generate the outputs of an STFT and manually visualise the output. This means you are totally in

---

[3]The spectrogram function has replaced the specgram function. specgram is still supported (without significant support of documentation, but is likely to be removed in future MATLAB versions).

[4]This generates the spectrogram shown in Figure 3.28.

[5]This generates the spectrogram shown in Figure 3.29.

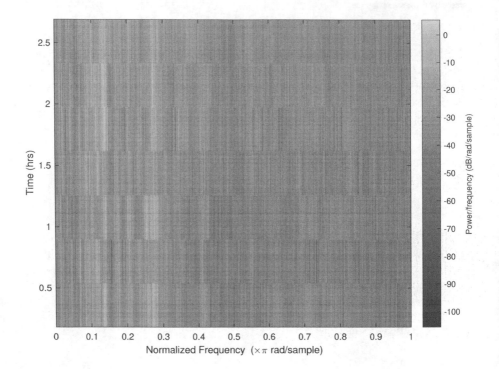

FIGURE 3.28: The spectrogram of handel.mat audio signal (using the default settings).

control of the visualisation parameters and not at the mercy of the sometimes arbitrary selection of default visualisation parameters of the `spectrogram` function itself.

Figure 3.30 shows a spectrogram of the end of an audio track called Windowlicker by The Aphex Twin. The composer of this piece has placed a picture of themselves (presumably Richard James, the real name of The Aphex Twin) within the music by inverting the spectrogram generation process and using a visual image as input. The following code shows how the image can be extracted from the audiofile using the `spectrogram` MATLAB function.

This figure (Figure 3.30) is created through the following MATLAB code:

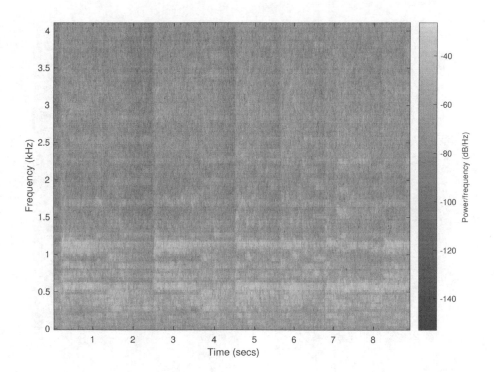

FIGURE 3.29: The spectrogram of handel.mat audio using the code given above.

```matlab
1  % Generate spectrogram of the Aphex Twin track "equation".
2  urlwrite('http://www.aphextwin.nu/visuals/hiddenfaces/
        equation9sec.wav', 'equation9sec.wav');
3  [Y, FS]=audioread('equation9sec.wav');
4  Y = mean(Y')';
5  F=logspace(log10(100),log10(22050),1024);
6  [y,f,t,p] = spectrogram(Y,1024,[],F,FS);
7  imagesc(t,f,10*log10(abs(p)));
8  axis xy;
9  xlabel('Time (s)', 'fontsize', 16);
10 ylabel('Frequency (Hz)', 'fontsize', 16);
11 set(gca,'Yscale','log');
12 colormap hsv;
```

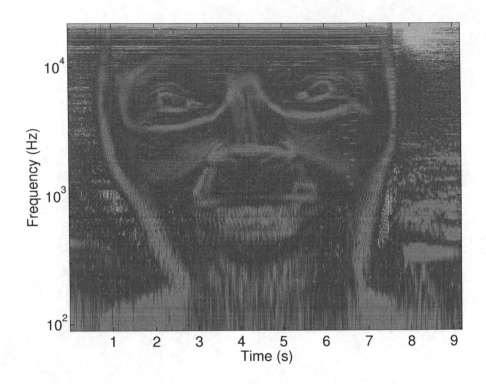

FIGURE 3.30: An example spectrogram of the audio track "Windowlicker" by the band The Aphex Twin.

3.18   Summary

- This chapter introduces the basics of frequency analysis for audio signals and their implementation using MATLAB.

- Topics introduced included

    - The four basic forms of the Fourier transform

    - An analysis of the DFT transform

    - The introduction of the FFT and its derivation

    - The overlap and add systems for perfect reconstruction including the COLA and WOLA methods and conditions for perfect reconstruction

    - Spectrograms

3.19   Exercises

Exercise 3.1

Make a spectrogram of the Aphex Twin track with decreased spectral and temporal resolution (so it looks blocky and subsampled in both time and frequency axes).

Exercise 3.2

Adjust the code in Listing 3.1 so only the phase is set to zero in the frequency range selected rather than the magnitude.

Bibliography

[1] J. Cooley and J. Tukey.   An algorithm for the machine calculation of complex fourier series. *Mathematics of computation*, 19(90):297–301, 1965.

4

Acoustics

CONTENTS

> To answer a popular question, if we define sound as travelling fluctuations in air pressure, then yes, a tree falling in the woods with no one nearby does indeed make a sound. If we define sound to be the electrical signal transmitted by the mechanisms of our inner ears to our brains, then no, that tree falling in the woods makes no sound.
>
> G. Lyons

4.1 Introduction

This chapter introduces some of the fundamentals of acoustics. Firstly, the wave equation is derived for both a vibrating string and a vibrating column of air. The solution to the wave equation for a vibrating column of air is then used as a basic model of speech production. Further examples of the wave equation are then investigated together with some special topics such as room reverberation.

Firstly the concept of simple harmonic motion (described in the first chapter) is revisited.

- Simple harmonic motion is the basic and most easily understood (mathematically) periodic oscillation.

- From simple harmonic motion derive the wave equation firstly for a string and then for a column of air.

- From the wave equation for a vibrating column of air a model for the formant frequencies of speech is derived.

4.1.1   The Wave Equation

In order to gain a key understanding of vibrations of natural objects within the field of acoustics, the definition of the wave equation is essential. Very significant complexity in the definition and solution of the wave equation is found for complex structures. We therefore firstly constrain the modelled vibration as a stretched string attached at two ends (e.g., a string on a guitar).

4.2   Wave Equation: Stretched String Attached at Both Ends*

In order to formulate the wave equation for a stretched string (anchored at both ends), a small section of the string is considered. Figure 4.1 shows such a small vibrating string section.

For this section of the string, the net vertical force is:

$$F_y = P\sin\theta_1 - P\sin\theta_2. \tag{4.1}$$

Which can be written as (Δ corresponding to the "change" within the parentheses):

$$F_y = P\Delta(\sin\theta). \tag{4.2}$$

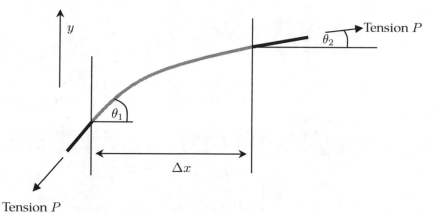

FIGURE 4.1: Small section of a vibrating string that leads to the formalisation of the wave equation for a vibrating string.

Box 4.2.1: Wave Equation Assumptions (Vibrating String)

The main assumptions made in this model are:

- Uniform linear density

- Negligible bending stiffness

- P large enough to rule out effects of gravity

- No dissipative forces

- Small angles

- x is the distance along the string

- t is the time

Given string density ρ (mass per unit length), Newton's 2nd law gives:

$$F_y = ma = \rho \Delta x \frac{\partial^2 y}{\partial t^2}. \tag{4.3}$$

Now equating (4.3) and (4.2) gives:

$$P\Delta\left(\sin\theta\right) = \rho \Delta x \frac{\partial^2 y}{\partial t^2}. \tag{4.4}$$

Rearranging gives:

$$(P/\rho)\frac{\Delta\left(\sin\theta\right)}{\Delta x} = \frac{\partial^2 y}{\partial t^2}. \tag{4.5}$$

For small values of θ the assumption can be made that $\sin(\theta) \approx \partial y/\partial x$. Together with assuming Δx tends to zero ($\Delta x \to 0$) this leads to the final result given in Box 4.2.2.

Box 4.2.2: The Wave Equation for a String Attached at Two Ends

$$\frac{\partial^2 y}{\partial t^2} = c^2 \frac{\partial^2 y}{\partial x^2}$$

- t is time

- x is distance along the string

- y is the vertical displacement of the string

- defining $c^2 = (P/\rho)$

- P is the string tension

- ρ is the string density (mass per unit length)

4.3   Wave Equation: Vibrating Stretched String Solution*

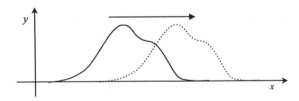

FIGURE 4.2: Moving wave on a string.

We begin the solution to the wave equation for a stretched string by considering a function f describing an arbitrary shape moving to the right on a string (as depicted in Figure 4.2). It is assumed that the shape of the function stays constant but the offset position changes in time. Such a function can be defined as:

$$y(x,t) = f(t - x/c),$$ (4.6)

where x, is the distance along the string, t is time and c is the speed of the wave on the string. This is also known as as d'Alembert's solution (see Box 4.3.1). This can easily be shown to be a solution to the wave equation (in Box 4.2.2) as follows:

$$f''(t - x/c) = \frac{\partial^2 y}{\partial t^2} = c^2 \frac{\partial^2 y}{\partial x^2} = c^2 f''(t - x/c).$$ (4.7)

This technique also works for a function g travelling in the opposite direction. Therefore the general solution is:

$$y(x,t) = f(t - x/c) + g(t + x/c).$$ (4.8)

Box 4.3.1: Jean-Baptiste le Rond d'Alembert

Jean-Baptiste le Rond d'Alembert (1717–1783) was a French mathematician, physicist, mechanician, philosopher and musical theorist. The wave equation is in some older texts referred to as d'Alembert's equation. d'Alembert's method for obtaining solutions to the wave equation for a vibrating string is named after him. (Public Domain Image).

4.4   Wave Equation: Reflection From a Fixed End*

In order to continue our analysis of the solutions of the wave equation of a vibrating string the condition where the string is fixed at one end is first imposed, i.e., $x = 0$ and $y = 0$ for all t. This condition is applied to (4.8) and leads to:

$$y(0,t) = f(t) + g(t) = 0. \tag{4.9}$$

Therefore:

$$f(t) = -g(t). \tag{4.10}$$

This therefore implies that given the above condition f is identical to g but with a reversed sign. Equation (4.8) therefore becomes:

$$y(x,t) = f(t - x/c) - f(t + x/c). \tag{4.11}$$

This makes intuitive sense and implies that the function shape of a travelling wave going to the right is matched to the same shape inverted travelling to the left as illustrated in Figure 4.3.

4.5   Wave Equation: String Vibration Fixed at Two Ends*

The next assumption is that the string is fixed at two ends, i.e., $y = 0$ for $x = 0$ and $x = L$. From this condition and (4.11):

$$y(L,t) = f(t - L/c) - f(t + L/c) = 0. \tag{4.12}$$

Adding L/c to both arguments gives:

$$f(t) = f(t + 2L/c). \tag{4.13}$$

This shows that f is periodic with a period of $2L/c$.

4.5.1   Solution to Wave Equation of a Vibrating String Fixed at Two Ends

If a solution (given $\omega_0 = \pi c/L$) is first guessed of the form:

$$y(x,t) = e^{j\omega_0 t}Y(x). \tag{4.14}$$

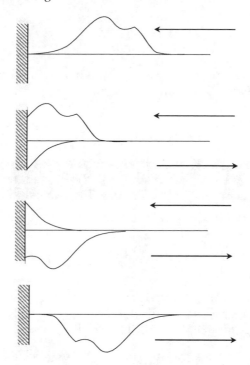

FIGURE 4.3: Reflection of travelling wave fixed at one end. A travelling wave travelling in the left direction is reflected exactly giving an identical wave travelling in the opposite direction. This results in an exact displacement y at all times t at $x = 0$ (with regard to (4.11)).

$$y(x,t) = e^{j\omega_0 t} Y(x)$$

$$\frac{\partial^2 y}{\partial t^2} = c^2 \frac{\partial^2 y}{\partial x^2}$$

$$\frac{\partial^2 Y(x)}{\partial x^2} = -(\pi/L)^2 Y(x)$$

The equation:

$$\frac{\partial^2 Y(x)}{\partial x^2} = -(\pi/L)^2 Y(x), \tag{4.15}$$

is now an intermediate solution to the wave equation for a vibrating string. However (4.15) is obviously just the equation of simple harmonic motion. The solution is therefore of the form (disregarding any amplitude weighting):

$$Y(x) = \sin(\pi x/L + \phi). \tag{4.16}$$

Enforcing the boundary conditions ($y = 0$ for $x = 0$ and $x = L$) leads to the conclusion that firstly $\phi = 0$ and therefore:

$$y(x,t) = e^{j\omega_0 t}\sin(\pi x/L). \tag{4.17}$$

Box 4.5.1: Solution to the Wave Equation String Vibration Fixed at Two Ends: Summary

- The solution to the wave equation for a string fixed at two ends ($x = 0$ and $x = L$) is:

$$y(x,t) = e^{j\omega_0 t}\sin(\pi x/L). \tag{4.18}$$

- All points on the string vibrate with a complex signal $e^{j\omega_0 t}$ with frequency $\omega_0 = \pi c/L$. The amplitude grows from the boundaries to the maximum at the middle according to $\sin(\pi x/L)$.

- The above solution can also be found using positive integer multiples of the phasor (k is a positive integer).:

$$y(x,t) = e^{jk\omega_0 t}\sin(k\pi x/L). \tag{4.19}$$

The solution to the wave equation given by (4.19) is illustrated in Figure 4.4. This figure shows the amplitude of the first six modes of the vibration of a string (i.e., the $\sin(k\pi x/L)$ part of (4.19)).

Figure 4.4 was produced using the following MATLAB code:

```
1   L = 1; % String length
2   x = 0:0.01:L;
3
4   for k = 1:6
5       subplot(6,1,k)
6       h(k) = plot(x,zeros(size(x)));
7       axis([0 1 -3/2 3/2])
8       set(h(k),'linewidth',2)
9       set(gca,'xtick',[],'ytick',[]);
10      ylabel(['k = ' num2str(k)]);
11      u = sin(k*pi*x/L);
12      set(h(k),'ydata',u);
13      set(gca,'fontsize', 14);
14  end
```

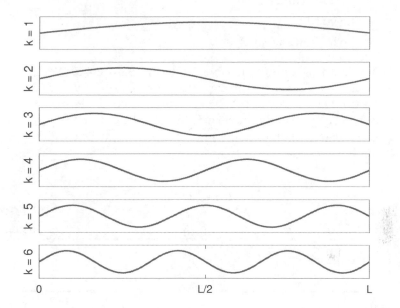

FIGURE 4.4: The first six modes of vibration $k = 1 \ldots 6$: Amplitude of equation (4.19).

4.6 Wave Equation: Vibrating Air Column

The analysis of a vibrating string using the wave equation is not only useful for the analysis of musical vibrations but it is instructive in the process that must now be followed to define and solve the wave equation for a vibrating column of air.

Furthermore, as the focus of this book is speech processing, the formalisation and solution for the vibration of air within constrained structures is the most applicable to the rest of the topics covered in this book. Initially the simplest case is considered: a vibrating column of air closed at one end and open at the other.

- For a column of air, the lateral displacement ξ of the air particles is synonymous with the vertical displacement y of a string within the model for a vibrating string.

- Excitation at one end of the column will produce a wave of varying air particle displacement and pressure.

- This model is useful to model

- Wind instruments
- The vocal tract
- The human ear canal

The considered column of air is illustrated in Figure 4.5. Waves within the column of air are formed through the compaction and rarification of air particles (higher and lower pressure) as illustrated in Figure 4.6. The lateral displacement of each of the air particles (shown by dots in this figure) denoted by ξ is the distance between any of each of these particles at any particular time and their resting state.

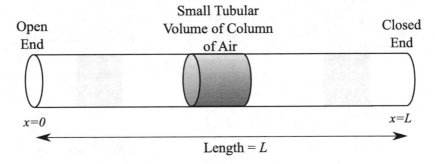

FIGURE 4.5: A column of air closed at one end.

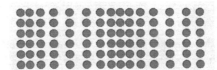

FIGURE 4.6: Propagation of a sound wave in a column of air (air particles shown as dots).

The model of the wave equation within a vibrating column of air is formulated by considering an elemental section of the air column as illustrated in Figure 4.7. The model parameters are defined within Figure 4.7 and Box 4.6.1.

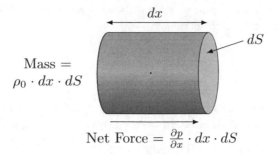

$$\text{Net Force} = \frac{\partial p}{\partial x} \cdot dx \cdot dS$$

FIGURE 4.7: Elemental volume of a vibrating column of air.

Box 4.6.1: Wave Equation Parameters (Vibrating Air Column)

The parameters of the wave equation formulated for a vibrating column of air (as illustrated in Figure 4.7) are:

- ξ = The lateral displacement of air particles
- ρ_0 = The density of the air at rest
- p = The pressure change at any place and time
- p_0 = Ambient pressure
- γ = Proportionality constant (coefficient of elasticity)
- Adiabatic compression is assumed
  - Any fractional change in pressure is proportional to the fractional change in density.

In order to generate the wave equation for a vibrating air column Newton's second law is equated to the formula denoting Adiabatic compression.

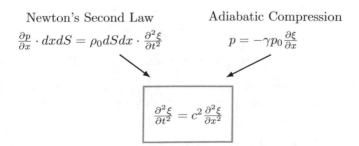

Newton's Second Law

$$\frac{\partial p}{\partial x} \cdot dx dS = \rho_0 dS dx \cdot \frac{\partial^2 \xi}{\partial t^2}$$

Adiabatic Compression

$$p = -\gamma p_0 \frac{\partial \xi}{\partial x}$$

$$\frac{\partial^2 \xi}{\partial t^2} = c^2 \frac{\partial^2 \xi}{\partial x^2}$$

Assuming that c is the speed of sound in air in the wave equation for a vibrating column of air and given the above formulation of the wave equation the value of c can be defined as:

$$c = \sqrt{\frac{\gamma p_0}{\rho_0}}. \tag{4.20}$$

The speed of sound in air is therefore dependent on the pressure and density of the air. This is also dependent on the temperature. At room temperature:

- Air has a coefficient of elasticity γ of $7/5 = 1.40$
- Air has a density ρ_0 of $1.2754\ kg/m^3$
- Air has a pressure of p_0 $101,000\ N/m^2$

The speed of sound predicted is therefore $\approx 330m/s$.

4.7 Wave Equation: Vibrating Column of Air (Open at One End) Solution

The solution to the wave equation for a vibrating column of air is initiated by considering a function f describing an arbitrary shape moving to the right. It is assumed that the shape of the function stays constant but the offset position changes in time. Such a function can be defined as:

$$\xi(x,t) = f(t - x/c). \tag{4.21}$$

Firstly, the open end $x = 0$ and the closed end $x = L$ are defined. The boundary conditions are as follows:

- At the open end $x = 0$, $\partial\xi/\partial x = 0$
  - The displacement amplitude must be at a maximum at the open end of the tube.
- At the closed end $x = L, \xi = 0$
  - The displacement amplitude must be zero at the closed end.

Similarly to the general solution to the wave equation for a vibrating string as given in (4.8) the solution here is also two arbitrary functions f and g as follows:

$$\xi(x,t) = f(t - x/c) + g(t + x/c) \tag{4.22}$$

where $f()$ is a right moving wave, $g()$ is a corresponding left moving wave.

4.7.1 Boundary Condition 1

At the open end ($x = 0$) the particle displacement ξ must be at a maximum. This condition can be defined as:

$$\text{at} \quad x = 0, \quad \frac{\partial \xi}{\partial x} = 0. \tag{4.23}$$

Combining this boundary condition and (4.22) leads to:

$$f'(t) = g'(t). \tag{4.24}$$

This implies that $f()$ and $g()$ differ by a constant. Ignoring this leads to:

$$\xi(x, t) = f(t - x/c) + f(t + x/c). \tag{4.25}$$

4.7.2 Boundary Condition 2

The condition associated with the closed end is that displacement must be zero, i.e., $\xi = 0$ when $x = L$. Therefore given this boundary condition and (4.25) leads to:

$$f(t - L/c) = -f(t + L/c). \tag{4.26}$$

Therefore, given both the conditions (and adding L/c to both function arguments) the following is obtained:

$$f(t) = -f(t + 2L/c). \tag{4.27}$$

This is almost the same result as for the vibrating string (4.13) apart from the sign. The function has a period of $4L/c$. The fundamental frequency is therefore defined as (in rad/sec):

$$\boxed{\omega_0 = \pi c/(2L)}$$

Guessing the form of the solution to the wave equation as:

$$\xi(x, t) = e^{j\omega_0 t} \Xi(x). \tag{4.28}$$

Substituting into the wave equation:

$$\frac{\partial^2 \Xi(x)}{\partial x^2} = -(\pi/2L)^2 \Xi(x). \tag{4.29}$$

This is the equation of simple harmonic motion. The solution is therefore of the form:

$$\Xi(x) = \sin\left(\frac{k\pi x}{2L} + \phi_1\right).\tag{4.30}$$

The solution to the wave equation gives (swapping cos for sin):

$$\Xi(x) = \cos\left(\frac{k\pi x}{2L} + \phi_2\right).\tag{4.31}$$

The boundary conditions give

- $\Xi'(0) = 0$ and $\Xi(L) = 0$
- This gives $-\sin(\phi_2) = 0$ and $\cos(\phi_2 + k\pi/2) = 0$
- We can therefore set ϕ_2 to 0
- k must therefore be odd if $\cos(k\pi/2) = 0$

The entire solution to the wave equation is thus of the form:

$$\xi(x,t) = e^{jk\omega_0 t}\cos\left(\frac{k\pi x}{2L}\right), \quad k = 1, 3, 5, \ldots\tag{4.32}$$

This equation has an interesting interpretation. The standing waves have odd harmonics (odd values of k). The resonant frequencies are $kc/4L$ for $k = 1, 3, 5, \ldots$. The fundamental is therefore $c/4L$ and the first two resonant frequencies are $3c/4L$ and $5c/4L$. For the vocal tract $L = 17.5cm$, $c = 330m/s$. The fundamental is therefore approximately 500Hz, first harmonic is 1500Hz and second harmonic is 2500Hz, etc. These are equivalent to vocal formants and gives an emphasis to the models that constrain the analysis of the most important elements of speech to be contained under 4kHz. The first three vibrational modes of a vibrating tube of air open at one end are illustrated in Figure 4.8.

4.7.3   Concatenated Tube Model

A single half-open vibrating tube model cannot effectively represent the time varying human vocal tract. A full three-dimensional model of the vocal tract is very complex and in most cases not necessary. A simple model of the vocal tract is just a concatenation of vibrating tubes of air as illustrated in Figure 4.9. The top of this figure illustrates the example width of the vocal tract (as a function of length along it). This is then modelled using a number of concatenated tube elements that closer approximates the actual vocal tract as the number increases.

The propagation and reflection of sound waves associated with the join between each pair of tube elements can be easily modelled as an all-pole filter. This, in turn, leads to the linear prediction speech analysis systems described in Section 9.4.

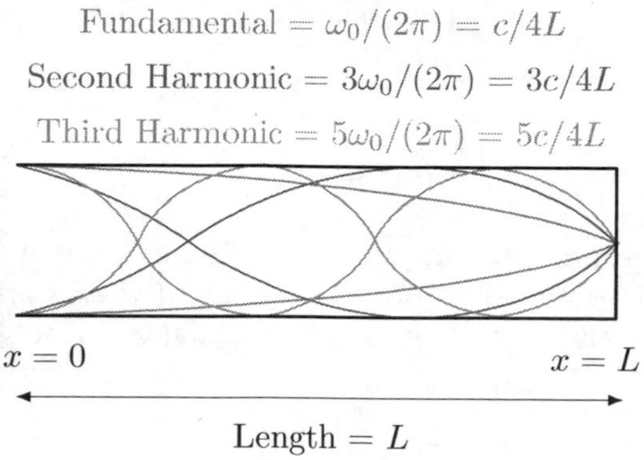

Fundamental = $\omega_0/(2\pi) = c/4L$
Second Harmonic = $3\omega_0/(2\pi) = 3c/4L$
Third Harmonic = $5\omega_0/(2\pi) = 5c/4L$
$x = 0$
$x = L$
Length = L

FIGURE 4.8: First three vibrational modes of a half open vibrating tube.

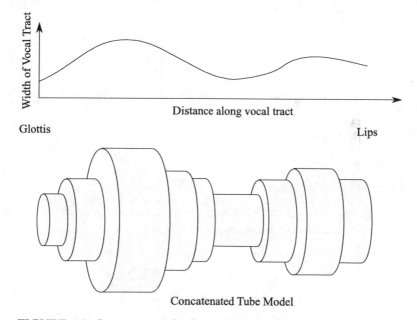

Width of Vocal Tract
Distance along vocal tract
Glottis
Lips
Concatenated Tube Model

FIGURE 4.9: Concatenated tube model for the human vocal tract.

4.8 Room Acoustics

Room acoustics is the study of the propagation of sound pressure throughout rooms.

4.8.1 Spherical Wave Equation

The spherical wave equation is equivalent to the 2D wave equation but for spherically propagated sound waves. It is important for understanding room acoustics. The three-dimensional wave equation can be expressed generically as:

$$\nabla^2 p = \frac{1}{c^2}\frac{\partial^2 p}{\partial t^2}, \tag{4.33}$$

where p is the sound pressure level, t is time and c is the speed of sound. Representing $\nabla^2 p$ in spherical polar coordinates (where r is the radial distance away from the source) and ignoring the angular contribution gives:

$$\nabla^2 p = \frac{\partial^2 p}{\partial r^2} + \frac{2}{r}\frac{\partial p}{\partial r}. \tag{4.34}$$

Combining (4.33) and (4.34) gives the spherical wave equation (radial r dependency only):

$$\frac{\partial^2 p}{\partial r^2} + \frac{2}{r}\frac{\partial p}{\partial r} = \frac{1}{c^2}\frac{\partial^2 p}{\partial t^2}. \tag{4.35}$$

The spherical wave equation describes pressure propagation of a point source ignoring

- Angular variations

- Directionality of the source

- Effect of boundaries

The radial solution (given pressure at time t and at radius r) is a complex sinusoid with amplitude P_0/r:

$$p(r,t) = \frac{P_0}{r}e^{i(\omega t - kr)}, \tag{4.36}$$

where

- P_0 is the amplitude of the sinusoid

- ω is the angular frequency of the wave

- k is the "wavenumber" (equal to $2\pi/(wavelength)$).

- r is the radial distance from the point source

The main conclusion from this solution is that **sound pressure is proportional to the inverse distance from a point source.**

4.8.2   Acoustics: Summary

- Simple harmonic motion is the most simple and easily understood mathematical representation of vibrations.

- The wave equation can be shown to be equivalent to a string under tension and a vibrating tube.

- The mathematical analysis of a tube fixed at one end shows that only odd harmonics can be present.

- It can be shown that the approximate fundamental of a human vocal tract is approximately 500Hz and the first two harmonics are at 1500Hz and 2500Hz.

- These frequencies are the equivalent of formants.

- The modelling of the human vocal tract can be achieved through the concatenation of tube models. This, in turn, can be modelled by a lattice filter.

- Sound pressure is proportional to the inverse distance from a point source (found as a solution to the three-dimensional wave equation).

4.9   Exercises

Exercise 4.1

Create MATLAB code to illustrate the amplitude of the solution to the wave equation for a string attached at both ends, i.e., factor $\sin(k\pi x/L)$ from (4.19) i.e. plot the amplitude from $x = 0$ to $x = L$ for an example length (e.g. $L = 1000$). Do this for values of k being 0.5,1,1.5,2,2.5 and 3 thus illustrating the boundary conditions are satisfied for k being a whole number (i.e., k is 1,2 and 3 in this case).

Exercise 4.2

Create MATLAB code to generate the amplitude of the solution to the wave equation for a half open tube, i.e., factor $\cos\left(\frac{k\pi x}{2L}\right)$ from (4.32), i.e., plot the amplitude from $x = 0$ to $x = L$ for an example length (e.g., $L = 1000$). Do this for values of k being 1,2,3,4 and 5 thus illustrating the boundary conditions are satisfied for k being a whole odd number (i.e., k is 1,3 and 5 in this case).

5

The Auditory System

CONTENTS

> The voice of the intellect is a soft
> one, but it does not rest until it
> has gained a hearing.
>
> Sigmund Freud

This chapter introduces the elements of the Human Auditory System (HAS) and then describes a number of auditory filter models based on the analysis of the HAS and peripheral system that will be used in subsequent chapters.

The applications explored within this book such as audio signal compression, speech coding and speech recognition require a fundamental understanding of the Human Auditory System (HAS). Specifically, effective auditory models for such applications need to be defined giving representations for the entire listening process, from the audio signal in free air, through the human auditory system to the resulting neural impulses within the brain. The physiology of hearing is analysed (the peripheral auditory system) together with the implications of the physical characteristics of the system and how this impacts on the signal transmitted to the listening centres of the brain. This analysis and a description of psychoacoustic experiments lead to defined frequency and temporal models of the HAS.

5.1 The Peripheral Auditory System

The physical elements of the human auditory system (the peripheral auditory system) are shown in Figure 5.1.

The Pinna: is commonly what is termed as the "ear" and is composed of a flap of skin together with a specifically shaped cartilage on each side of the head. The pinna forms what is known as the "outer ear" together with the ear canal. The pinna is shaped specifically to direct sound waves into the ear canal and toward the rest of the periphery. It plays a key role in providing binaural and mono-aural sound location cues due to the significant differences in perceived loudness and timbre of a sound depending on the source's location and direction.

The Ear Canal: is a hollow tube linking the tympanic membrane to the pinna. Although the physiology of each person's ear canal can vary significantly, the dimensions of a typical ear canal are such that a resonance at approximately 2–5kHz significantly emphasises this range of frequencies.

The Tympanic Membrane: is a membrane stretched across the end of the ear canal (it is also known as the eardrum). It is able to vibrate across the range of human hearing frequencies and transmit the vibration via the ossicles to the cochlea via the oval window.

The Ossicles (Malleus, Incus and Stapes): are three very small bones connected to the tympanic membrane that are caused to vibrate by the tympanic membrane and amplify the vibration before transmitting it to the oval window of the cochlea. They convey the air-transmitted sound pressure waves causing the tympanic membrane to transmit vibrations to the fluid-transmitted sound pressure waves within the cochlea.

The Oval Window: is a membrane on the entrance of the cochlea. The ossicles vibrate the oval window transmitting the sound energy to the basilar membrane and the fluid within the cochlea.

The Basilar Membrane (BM): is a bone that partitions the spiral shaped cochlea. It acts as a logarithmic spectrum analyser, physically dividing and isolating the input vibration frequency components to different physical locations along its length. Auditory neurons are connected to different types of hair cells (inner and outer) in turn attached to the basilar membrane. The vibration of the basilar membrane is therefore transmitted to higher centres of the auditory neural pathway via these hair cells.

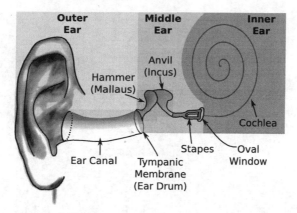

FIGURE 5.1: The peripheral auditory system.

5.1.1 The Basilar Membrane (BM)

The basilar membrane is narrow and stiff at the base and near the apex it is wider and less stiff. The peak amplitude of vibration along the BM is dependent on the frequency of the sound listened to. Higher frequencies have a peak amplitude near the base, lower frequencies have peak amplitude near the apex. The physics of the BM and the cochlea can, therefore, be considered as a set of bandpass filters. BM motion is transmitted to the stereocilia. The stereocilia are the hair bundles attached to the top of each hair cell. This leads to the firings of the peripheral auditory neurons. The physical response of the BM according to distance and frequency is illustrated in Figures 5.4 and 5.5.

Figure 5.2 shows a cross-section within the cochlea taken from Gray's anatomy [7]. This figure shows the basilar membrane as a horizontal line. Above the BM, four rows of auditory hair cells are also shown. These four rows are separated into one single row of auditory cells called the inner hair cells and three rows of auditory hair cells called the outer hair cells. These different hair cells and the auditory mechanisms within the cochlea are able to function over a very large dynamic range of approximately 100dBs. [1]

The human cochlea has approximately 3,500 inner hair cells and 12,000 outer hair cells at the time of birth [2].

5.1.2 Auditory Pathways

Figure 5.3 shows the auditory pathways from the cochlea to the higher levels of auditory processing centres in the brain. The key points to notice in this diagram is that auditory pathways go in *both* directions, i.e., from cochlea to the auditory cortex and from the auditory cortex to the cochlea. The charac-

---

[1]The choice of 16bits to represent samples on an audio CD can be interpreted within this context. Given the SNR per bit given by (2.46) the SNR of an audio CD is approximately $16 \times 6 = 96dB$ which is close to the 100dBs dynamic range of human hearing.

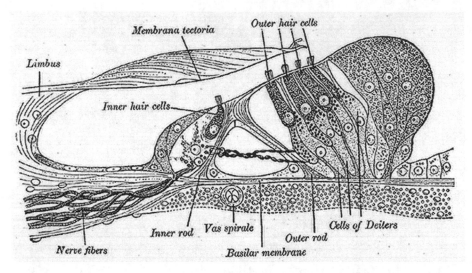

FIGURE 5.2: Organ of Corti: Taken from Gray's Anatomy plate 913 [7]. (Public Domain Image).

teristics of hearing discussed later within this chapter should be considered in the context of these forward and backward connections. Additionally, the left and right auditory processing chains are linked. These links give aid to balance in some cases and are the mechanism through which binaural three-dimensional localisation is assumed to be enacted.

5.2 Auditory Filters: Introduction

The frequency analysis and decomposition of an audio signal is of key importance within many of the applications discussed within this book (for example: compression, speech recognition, speech coding, etc.). As many of the most effective methods within these applications have attempted to mimic aspects of the human audio system, a fundamental appreciation of the way that humans (and mammals) perceive audio frequencies is essential to understand the applications that follow. For speech analysis applications such as recognition tasks, it is often most effective to isolate and characterise the spectral envelope of a small temporal section of audio. This is because it has been found in the vast majority of cases that the semantic meaning of speech is not dependent or contained within the pitch or excitation component of the speech signal. Methods to characterise this short-term spectral envelope include the Short Term Fourier Transform (STFT-most often implemented using windowed FFT) and Linear Predictive Coding (LPC). Historically such a

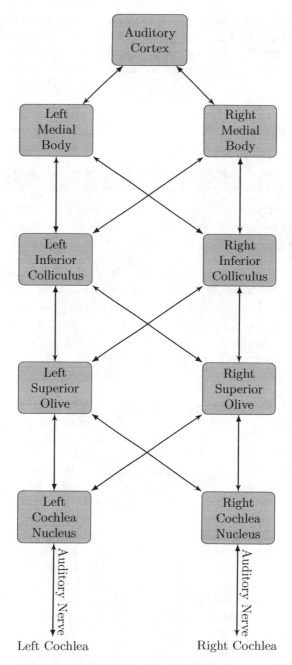

FIGURE 5.3: Auditory pathways linking the left and right ear to the auditory cortex (forward and backward connections).

characterisation and analysis have been achieved using a bank of overlapping bandpass filters. This structure of a set of bandpass filters (known as critical bands) is the current consensus view of the human frequency analysis/hearing mechanism and is explored in further detail in this chapter.

5.3   Critical Bands

> **Box 5.3.1: Definition of Critical Bands**
>
> Auditory critical bands can be defined as the range of audio frequencies within which a masking tone will interfere with the perception of another tone through various mechanisms of auditory masking. The understanding and use of critical bands is key for the implementation of perceptual frequency-based audio processing as used within applications such as wideband compression (e.g., MP3 compression).

In order to understand the perceptual frequency audio domain, it is necessary to understand the so-called frequency-to-place transformation model. This model theorises that each audio frequency relates to a specific place within the human cochlear and specifically a position along the basilar membrane (BM).

An audio wave entering into the ear has its vibrational characteristics transmitted to the oval window via the vibration of the tympanic membrane and the ossicles (shown in Figure 5.1). The vibration of the oval window then induces a vibration of the coiled BM (also shown in Figure 5.1 as the coiled spiral structure). For a pure tone, the maximum magnitude vibration of the BM is located at a specific physical position along the BM (the position contiguously ordered in terms of the fundamental frequency of the tone). As the distance from the oval window increases, the width of the BM increases and conversely its stiffness decreases. The effect of these physical characteristics is that higher frequencies have a peak amplitude along the BM closer to the oval window and stapes (and conversely there is a maximum further from the oval window toward the BM apex for lower frequencies). When approaching this position of maximum displacement the wave slows, reaches its peak (at the given location) and then rapidly decays along the remaining length of the BM.

An illustration of a single response along the BM for a tone at 200Hz is shown in Figure 5.4. The four graphs in this figure show the BM displacement amplitude at four contiguous times (with the amplitude response shown as a dotted line). The amplitude response along the BM to eight pure tones of frequency centres 25, 50, 100, 200, 400, 800 and 1600Hz are shown in

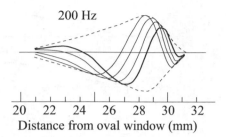

FIGURE 5.4: The displacement response of the BM for a 200Hz pure tone (x-axis is the distance along the BM and the y-axis is the displacement). The four graphs show the displacement at four contiguous times. Reprinted with permission: From G. Von Békésy. *Experiments in Hearing*. McGraw-Hill series in Psychology. McGraw-Hill, 1960. Copyright 1960, Acoustic Society of America [19].

Figure 5.5. [2] This shows that for a pure tone, the maximum amplitude of the BM vibration is located at a specific position corresponding to a given frequency. For a more complex tone, the component harmonics physically excite set positions along the BM according to the harmonic structure (and magnitude of the sound). This physical frequency separation mechanism, therefore, represents a frequency-to-place transformation similar in action to a bank of overlapping bandpass filters or similar frequency decomposition from a signal processing perspective.

Additionally, as illustrated in Figures 5.4 and 5.5, each of the pass-bands of the physical "BM filter" are firstly non-linear, secondly asymmetric and lastly have a non-uniform bandwidth with the bandwidth increasing with increasing frequency. These BM pass-bands and their bandwidths can be interpreted as "critical bands" and the "critical bandwidths," respectively. It has been shown that if frequency components of an audio signal are sufficiently separated they will be separately coded by the hearing system with a set of auditory filters.

However, it should be noted that the perception of pitch is much more complicated than this simplistic "place theory" (i.e., the theory that the perception of pitch corresponds to a set place on the BM). Furthermore, the "place theory" gives a good illustration of how the HAS has a variable bandwidth of influence with frequency. This general principle has been backed up by a multitude of experimental tests and measurements that not only have attempted to generate models of the bandwidth of the auditory filters but their actual shape (see below).

---

[2]These graphs were taken from work by Von Békésy [19] where he measured BM displacements of a human cadaver.

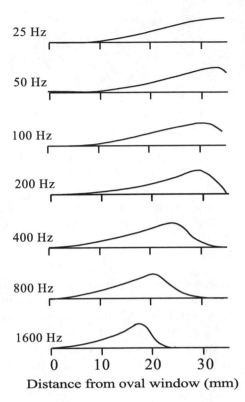

FIGURE 5.5: Amplitude displacements of the basilar membrane for eight pure tones. Reprinted with permission. From: G. Von Békésy. *Experiments in Hearing*. McGraw-Hill series in Psychology. McGraw-Hill, 1960. Copyright 1960, Acoustic Society of America [19].

5.4   Critical Band Models

5.4.1   Measurement of Critical Bandwidths by Fletcher*

Fletcher [4] attempted to measure the width of the critical bands of the auditory filters using a test tone and co-located wideband noise (as illustrated in Figure 5.6). The threshold of perception for a single test tone was measured in the presence of wideband noise. The noise was of a set bandwidth centred on the test tone. This bandwidth was varied as illustrated in Figure 5.7 resulting in a threshold of perception for the test tone for each noise bandwidth. The results showed that as the wideband noise bandwidth increases, so does the intensity (threshold of perception) of the co-located tone up to a certain point (of Δf). Past this point, the threshold does not increase and stays steady at the same value. The whole experiment was repeated for different tone frequencies giving results such as those shown in Figure 5.8. This figure is actually from a repeat of similar experiments by Schooneveldt and Moore [6] in 1987. This figure clearly shows the increasing threshold as the noise bandwidth increases up to the critical bandwidth of 400Hz and then remains constant after that.

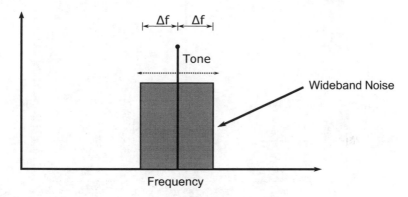

FIGURE 5.6: Wideband noise "critical bandwidth" experiments by Fletcher [4].

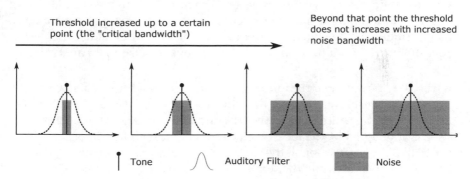

FIGURE 5.7: Illustration of the experiments by Fletcher [4].

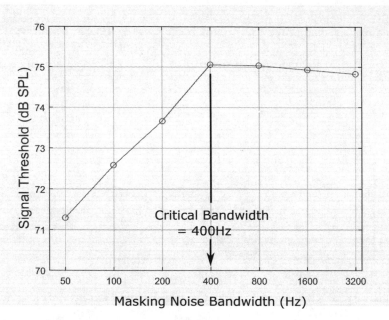

FIGURE 5.8: Tone threshold results in co-located wideband noise (for a 2kHz test tone). Reprinted with permission. From: G.P. Schooneveldt and B.C.J. Moore. Comodulation masking release for various monaural and binaural combinations of the signal, on frequency, and flanking bands. *J Acoust Soc Am.*, 1(85):262-272, 1987. Copyright 1987, Acoustic Society of America [6].

Given the "power spectrum model of hearing" the critical bandwidth phenomena is a result of auditory detection in terms of a signal-to-noise (SNR) criterion [11]. Specifically, as the bandwidth of the noise starts to increase very little is attenuated by the local auditory filter and therefore in order to retain the SNR for a given listener, the threshold increases. Conversely, as the noise bandwidth increases into the stop bands of the filter, the amount of perceived noise will remain constant and therefore the threshold remains constant.

5.4.2 The Bark Scale

A key and often used representation of the critical band domain is the "Bark Scale". The Bark scale is most often defined as a table of frequency values (see Table 5.1). These values define the centre frequencies and frequency ranges (and thus critical bandwidths) of critical bands. They have been taken from the paper by Zwicker [21] that originally defined the Bark scale. [3]

Although the Bark scale was defined as the table shown in Table 5.1, several analytic expressions for the conversion between the frequency (in Hz)

---

[3]The Bark scale was named after Barkhausen, a notable auditory scientist who studied the perception of loudness in the early part of the 20th century.

TABLE 5.1: Bark scale: band number, centre frequency and frequency ranges

Bark	Cntr Freq(Hz)	Freq Rng(Hz)	Bark	Cntr Freq(Hz)	Freq Rng(Hz)
1	50	20–100	13	1850	1720–2000
2	150	100–200	14	2150	2000–2320
3	250	200–300	15	2500	2320–2700
4	350	300–400	16	2900	2700–3150
5	450	400–510	17	3400	3150–3700
6	570	510–630	18	4000	3700–4400
7	700	630–770	19	4800	4400–5300
8	840	770–920	20	5800	5300–6400
9	1000	920–1080	21	7000	6400–7700
10	1170	1080–1270	22	8500	7700–9500
11	1370	1270–1480	23	10500	9500–12,000
12	1600	1480–1720	24	13500	12,000–15,500

and the Bark scale have been proposed. The most commonly used is that defined by Zwicker [22] shown in (5.1). It should be noted that this equation (and the other related models) take the value of frequency (f) to be the upper frequency of the range for the critical band rate in Barks (the right value in the right column in Table 5.1).

$$\text{Bark}(f) = 13 \cdot \tan^{-1}(0.00076 \cdot f) + 3.5 \cdot \tan^{-1}\left((f/7500)^2\right) \qquad (5.1)$$

Listing 5.1: MATLAB code to create Figure 5.9

```
1  f = 20:20000;
2  bark = [100 200 300 400 510 630 770 920 1080 1270 1480 1720 2000
          2320 2700 3150 3700 4400 5300 6400 7700 9500 12000 15500];
3  plot(barkCenters, [1:24],'k+'); hold on;
4  zf2 = 13*atan(0.00076*f)+3.5*atan((f/7500).^2);
5  plot(f,zf2,'r');
```

Figure 5.9 shows these values from table 5.1 in conjunction with the approximation Equation (5.1).

Furthermore, the bandwidths of each Bark filter can be approximated by the following function defined by Zwicker and Terhardt [22]:

$$\text{Bark}(f) = 25 + 75\left(1 + 1.4\,(f/1000)^2\right)^{0.69}. \qquad (5.2)$$

This equation agrees with the tabulated bandwidth values given in table 5.1 from DC to 15kHz within 10%.

There have been further updates and corrections to the Bark formulae (notably by Traunmuller [18], Schoeder et al. [15] and Wang et al. [20]). One of the most recently used approximations to the Bark scale is provided by Wang et al. [20]:

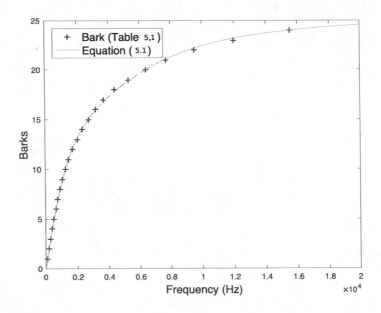

FIGURE 5.9: Comparison of Bark scale values from Table 5.1 and (5.1).

$$\text{Bark}(f) = 6\sinh^{-1}(f/600). \tag{5.3}$$

This formula is used within the PLP-RASTA perceptual feature extraction process [8] as described in Chapter 9.

5.4.3   Mel Perceptual Scale

Another popular perceptual frequency scale is the Mel (short for **Mel**ody) scale. It is derived from the original work by Stevens et al. [16] in 1937. Similarly to the Bark scale, there are numerous formulae that are used to convert between the Mel perceptual scale and linear frequencies (in Hz). A popular and common formula to convert from frequency to Mels is:

$$\text{Mel}(f) = 2595 \log_{10}\left(1 + \frac{f}{700}\right). \tag{5.4}$$

It is used with the MFCC feature extraction methods described in Section 9.6 as implemented within the popular speech toolkit HTK [3].

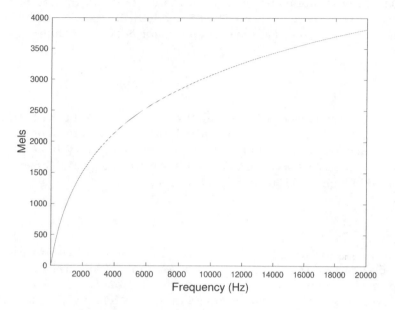

FIGURE 5.10: Mel scale as shown given by (5.4) generated by code in Listing 5.2.

Listing 5.2: MATLAB code to create Mel scale Figure 5.10

```
1  f = 20:20000;
2  mel = 2595*log10(1+f/700);
3  plot(f,mel,'r');
```

5.4.4   Measurement of Auditory Filter Shapes*

Many different experiments have been conducted into obtaining the shape of the auditory filters [9, 11, 12, 13]. A significant initial attempt to characterise and measure the shape of auditory filters was made by Patterson [12]. Patterson conducted a number of experiments using a fixed test tone and highpass or lowpass wideband noise of variable bandwidth. This masking signal was chosen to be wideband noise, as sinusoids and narrowband noise signals were found to give temporal artefacts (e.g., beating with sinusoids) [11]. Figure 5.11 shows an illustration of such experiments with lowpass noise (taken from Patterson [12]). Firstly, the aim of these experiments was to find the threshold of detection for a test tone (shown as the single vertical line in the middle of the figure). The lowpass noise was then varied in bandwidth (from DC to a given cutoff frequency). This was reflected in the Δf parameter shown in the figure that gives the distance between the edge of the noise and the tone. Using what is known as the "power spectrum model", the area of overlap between noise and filter (i.e., the integration of the product of the noise and filter functions) is directly related to the threshold of detection for the test tone. This model of the overlap was formulated by Patterson [12] as follows:

$$P_s = K \int_0^\infty |H(f)|^2 N(f) df, \tag{5.5}$$

where P_s is the power of the signal at the threshold of perception, K is a constant, $H(f)$ is the transfer function of the auditory filter response and $N(f)$ is the wideband noise masking power spectrum. Obviously, if the wideband noise is an ideal lowpass noise filter with a sharp cut-off at frequency W (as depicted in Figure 5.11) then (5.5) simplifies to:

$$P_s = K N_0 \int_0^W |H(f)|^2 df, \tag{5.6}$$

where N_0 is the value of the lowpass noise function $N(f)$ below the cutoff. The spectrum level of the noise was kept constant at 40dB in all of Patterson's experiments while the cutoff was varied and the threshold value of the test tone measured. The results are shown in Figures 5.11 and 5.12 for lowpass and highpass noise, respectively. Five graphs are shown in each of the figures. These graphs show the threshold experiment results for the test tone frequencies of {0.5, 1, 2, 4 and 8} kHz. Four key conclusions from these experiments were:

- The bandwidths of the filters increase with the central frequency of the test tone.

- The filters are not "constant Q", i.e., the ratio of the filter bandwidths to centre frequency is not constant (it increases with frequency).

- The filters are nearly symmetric.

- The attenuation rate of the filter skirts is high.

A seventh order polynomial function was fitted to each of these curves and differentiated analytically. A simplified expression of these derivatives was used to obtain a simple definition of the auditory filter given in Equation (5.7):

$$|H(f)|^2 = \frac{1}{\left[((f - f_0)/\alpha)^2 + 1 \right]^2}, \tag{5.7}$$

where f_0 is the frequency of the filter centre and α is the selectivity of the filter, i.e., as its value increases so does the width of the filter. Figure 5.13 shows a graph of $|H(f)|^2$ versus frequency for various values of α. This figure shows the symmetric bell-shaped nature of the filter.

1.29α is the 3-dB bandwidth (BW) of the filter. Using the integral of this filter

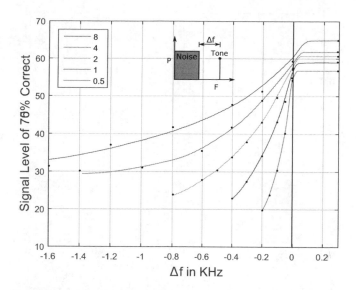

FIGURE 5.11: Result of noise threshold experiments for lowpass, wideband noise. Reprinted with permission. From: R.D. Patterson. Auditory filter shape. *J Acoust Soc Am.*, pages 802-809, 1974. Copyright 1974, Acoustic Society of America [12].

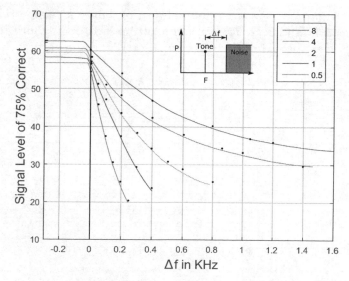

FIGURE 5.12: Result of noise threshold experiments for highpass, wideband noise. Reprinted with permission. From: R.D. Patterson. Auditory filter shape. *J Acoust Soc Am.*, pages 802-809, 1974. Copyright 1974, Acoustic Society of America [12].

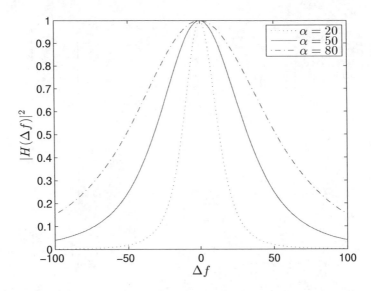

FIGURE 5.13: Symmetric auditory filter model proposed by Patterson [12] (using three values of α {20, 50, 80}) showing that the bandwidth of the filter increases as α increases (from Equation (5.7)). Generated from Listing 5.3.

function, the resulting curves fitted almost exactly with the curves shown in Figures 5.11 and 5.12. The variation of the 3-dB bandwidth (BW) with frequency was approximated by the equation (Patterson [12]). [4]:

$$10\log_{10}(BW) = 8.34\log_{10}(f) - 7.37. \tag{5.8}$$

---

[4]This relation therefore directly relates the centre frequency of the filter to the spreading coefficient α given that the 3-dB BW = 1.29α.

Listing 5.3: MATLAB code to create Figure 5.13

```
1  df = -100:100;
2  alpha = 20;
3  p = 1./ (((((df)/alpha).^2)+1).^2;
4  plot(df,p,':');
5  hold on;
6  alpha = 50;
7  p = 1./ (((((df)/alpha).^2)+1).^2;
8  plot(df,p,'-');
9  alpha = 80;
10 p = 1./ (((((df)/alpha).^2)+1).^2;
11 plot(df,p,'-.');
```

5.4.5 Notched Noise Experiments*

It was recognised by Patterson [13] that trying to obtain the shape of the auditory filter from Equation (5.6) made the possibly invalid assumption that the auditory filter was centred at the listening frequency (the phenomena of an auditory filter not being centred at the listening frequency is known as "off centre listening").

In order to generate the shape of auditory filters without this limiting assumption, "notched noise experiments" were used. For a fixed shape of the auditory filter $H(f)$ transfer function and lowpass noise from 0 to $f_c - \Delta f$ and highpass noise from $f_c + \Delta f$ to ∞ Equation (5.5) becomes:

$$P_s = KN_0 \int_0^{f_c - \Delta f} |H(f)|^2 \, df + KN_0 \int_{f_c + \Delta f}^{\infty} |H(f)|^2 \, df, \qquad (5.9)$$

where K is a constant, N_0 is the noise level, the masking noise has a notch of width $2\Delta f$ centred at f_c. P_s therefore represents the intersection area between the noise and auditory filter (i.e., the shaded intersection region for both auditory filters in the bottom illustration in Figure 5.14).

Notched noise experiments were created to overcome the assumption that the analysed auditory filter was centred at the frequency of the test (test tone or other method). Figure 5.14 illustrates that if the auditory filter is not centred at the test tone then the intersection of the filter and the notched-noise (i.e., the intersection area with the high and lowpass parts of the noise) gives a value P_s in (5.9) that is approximately constant. This can never be the case for the separate high or lowpass experiments illustrated in Figures 5.11 and 5.12. Large differences in bandwidth and filter shape results using notched noise experiments compared to previously more naive methods indicate that the "off centre listening" effect is significant [11].

The result of using notched noise experiments led to an update of Equation (5.8) by Patterson [13]:

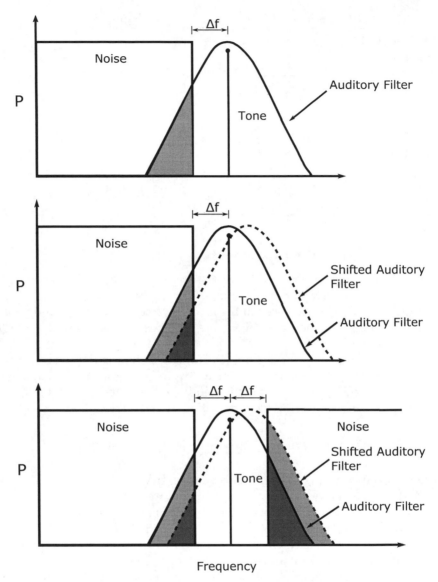

FIGURE 5.14: Justification of notched noise experiments. Top: Single lowpass noise experiment (similar to Figure 5.11). Middle: In such experiments, if the filter is not centred around the test tone, the shaded area of the interface between the filter and noise will not lead to an accurate model. Bottom: For such "off centre listening" the summation of the shaded regions is approximately constant. Redrawn with permission. From: R.D. Patterson. Auditory filter shapes derived with noise stimuli. *J Acoust Soc Am.*, pages 640-654, 1976. [13].

$$10\log_{10}(BW) = 7.91\log_{10}(f) - 2.71. \tag{5.10}$$

i.e., the variation of bandwidth with frequency was very close to the previous model (5.8), but with the actual bandwidths being comparatively large.

5.4.6 RoEx Filters*

The notched filter experiments by Patterson [13] led to a significant update of the measured auditory filter bandwidths (i.e., from (5.8) to (5.10)). However subsequent experiments by Moore and Glasnery showed that a more accurate model of the auditory filter can be approximated by a **Ro**unded **Ex**ponential filter (defined as a **RoEx** filter). A single parameter RoEx filter can be defined as follows:

$$W(g) = (1 + pg)e^{-pg} \tag{5.11}$$

where

- p is the single controlling parameter
- f_0 is the filter's centre frequency
- f is the independent frequency variable
- g = the normalised frequency = $|f - f_0|/f_0$

Figure 5.15 shows three RoEx filters at f_0= {500, 1000 and 2000}Hz calculated from Equation (5.11).

In order to improve the accuracy of the RoEx filter model to fit experimental data a more flexible version with two controlling parameters has been used (RoEx (p,r)) defined as follows:

$$W(g) = (1 - r)(1 + pg)e^{-pg} + r. \tag{5.12}$$

The RoEx filters are symmetric and more recent notched noise experiments have shown that there is significant asymmetry of the auditory filter as the centre frequency increases (the shape of the filter is more and more skewed as the centre frequency increases). Additional flexibility is created for asymmetric modelling by using separate RoEx filters for the filter regions above and below the central frequency.

5.4.7 Equivalent Rectangular Bandwidths: ERBs*

The equivalent rectangular bandwidth is defined as the bandwidth of a square filter co-centred with the actual auditory filter but with equal height and total area (i.e., they both have the same energy). This definition is illustrated in Figure 5.16. This definition of the ERB and various forms of RoEx filter

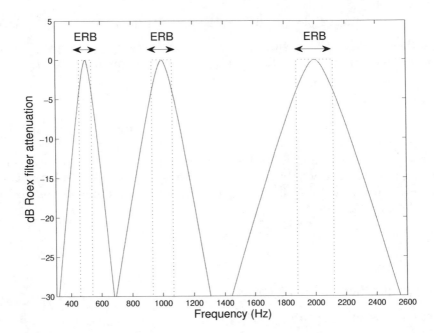

FIGURE 5.15: Three RoEx filters with Equivalent Rectangular Bandwidths.

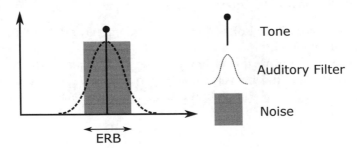

FIGURE 5.16: Equivalent Rectangular Bandwith (ERB)

together with a large number of notched noise experiments have generated a number of ERB scales. Firstly Equation (5.13) was generated from the fitting of notched noise experiments by Moore and Glasberg [10]:

$$\text{ERB}(f) = 6.23 \cdot (f/1000)^2 + 93.39 \cdot (f/1000) + 28.52. \qquad (5.13)$$

Where $\text{ERB}(f)$ is the ERB bandwidth at frequency f. This was updated in 1990 (by Glasberg and Moore [5]) to the currently preferred (and simpler) formula:

$$\text{ERB}(f) = 24.7 \cdot (4.37 \cdot f/1000 + 1). \qquad (5.14)$$

5.4.8 Octave Band Critical Bandwidth Filters*

The sophisticated critical bandwidth models above can be compared to simplistic octave-based critical bandwidth models. These models assume that f_l and f_u are the lower frequency and upper frequencies of a critical band respectively defined in the following relationship:

$$\frac{f_u}{f_l} = 2^k \tag{5.15}$$

where $k=1$ for octave band filters and $k=1/3$ for third-octave band filters.

Given some rudimentary manipulation leads to a simple definition of the bandwidths in each case:

Octave bands: bandwidth = 0.707×centre frequency

1/3 Octave bands: bandwidth = 0.232×centre frequency

5.4.9 Comparison of Critical Bandwidth Models as a Function of Frequency*

As a conclusion to these sections describing critical bandwidth models, they are compared graphically. MATLAB code is given in code Listings 5.4 and 5.5 for the following comparison figures. Firstly Figure 5.17 shows the Bark centre frequencies and bandwidths as tabulated in Table 5.1. The continuous approximation to these data points is given by Equation (5.2) defined by Zwicker and Terhardt [22] and also illustrated in the graph. Equations (5.8) and (5.10) also are illustrated within this graph. This shows that the rate increase (i.e., slope) of these two models is approximately the same.

Figure 5.18 also shows the Bark centre frequencies and bandwidths. Alongside these repeated values and graphs are the results from the ERB formulas; Equation (5.13) and Equation (5.14). These graphs show that the models are firstly very similar to each other but also that these models (using notched noise and non-symmetric RoEx filter matching) show that the critical bandwidths significantly reduce below the "knee" of the Bark scale (i.e., below about 100Hz, the Bark scale bandwidths stay constant at approximately 20Hz, whereas the ERB models' bandwidths continue getting smaller as the frequency decreases). **From the studies considered, Equation (5.14) provides the best and most up to date approximation to the change of critical bandwidth with frequency.**

Listing 5.4: MATLAB code to create Figure 5.17

```
1  f = 20:20000;
2  barkCenters = [150 250 350 450 570 700 840 1000 1170 1370 1600
       1850 ...
3     2150 2500 2900 3400 4000 4800 5800 7000 8500 10500 13500];
4  barkBW = [100 100 100 110 120 140 150 160 190 210 240 280 320
       ...
5     380 450 550 700 900 1100 1300 1800 2500 3500];
6  loglog(barkCenters, barkBW,'k+'); hold on;
7  terhar = 25+(75*(1+1.4.*(f./1000).^2).^0.69);
8  loglog(f,terhar,'k');
9  of = f*0.707; %octave filters
10 tof = f*0.231; %third-octave filters
11 loglog(f,of,'r');
12 loglog(f,tof,'b');
13 BW1 = 10.^((7.91*log10(f)-2.71)/10);
14 BW2 = 10.^((8.34*log10(f)-7.37)/10);
15 loglog(f,BW1,'g');
16 loglog(f,BW2,'r:');
```

Listing 5.5: MATLAB code to create Figure 5.18

```
1  f = 20:20000;
2  barkCenters = [150 250 350 450 570 700 840 1000 1170 1370 1600
       1850 ...
3     2150 2500 2900 3400 4000 4800 5800 7000 8500 10500 13500];
4  barkBW = [100 100 100 110 120 140 150 160 190 210 240 280 320
       ...
5     380 450 550 700 900 1100 1300 1800 2500 3500];
6  loglog(barkCenters, barkBW,'k+'); hold on;
7  terhar = 25+(75*(1+1.4.*(f./1000).^2).^0.69);
8  ERB1=24.7*(4.37*(f./1000)+1);
9  ERB2=6.23*(f/1000).^2+93.39.*(f./1000)+28.52;
10 loglog(f,terhar,'k');
11 loglog(f,ERB1,'r');
12 loglog(f,ERB2);
```

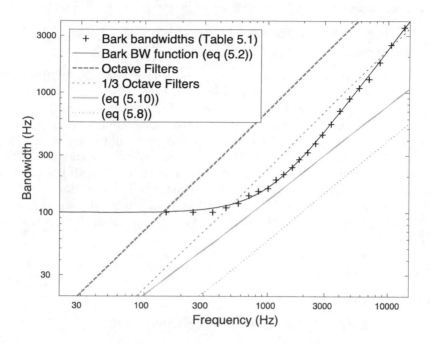

FIGURE 5.17: Comparison of Critical Bandwidth Models 1.

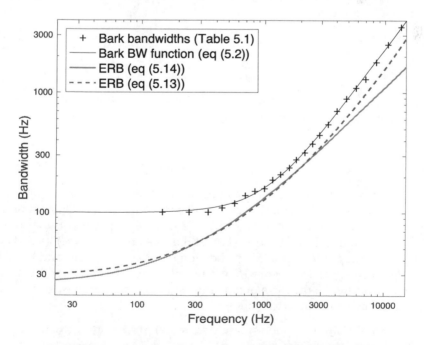

FIGURE 5.18: Comparison of Critical Bandwidth Models 2.

5.5 Relationship Between Filter Shape and the Masking Functions Used for Compression

The masking curves associated with the various forms of masking described within Chapter 7 have considerably wider high-frequency skirts and narrower lower frequency skirts (and in fact opposite to the asymmetric auditory filters measured by Moore and other authors). This is explained by the fact that such masking is the cumulative effect of the response of numerous auditory filters. This effect is illustrated within Figure 5.19. At any listening frequency, the response of the local auditory filter is its intersection with the test tone impulse. As illustrated by the figure, this leads to a wider masking response at frequencies higher than the masking tone. Such skewed masking curves are illustrated by the frequency masking models shown in Figures 7.10 and 7.11.

5.6 Properties of the Auditory Neurons*

In order to analyse the auditory system as a whole it is key to analyse the properties of single auditory neurons. The key characteristics of auditory neurons to analyse are:

- Adaptation

- Tuning

- Synchrony

- Non-linearity (including masking)

5.6.1 Adaptation of the Auditory Nerve

As in common with many sensory neurons, auditory neurons exhibit adaptation behaviour. An example behaviour of an auditory nerve is illustrated in Figure 5.20. Figure 5.20 shows the result of a series of experiments to produce what is known as a Post-Stimulus Threshold Histogram (PSTH). A PSTH is generated through the repeated application of a tone to the ear and the time taken for each spike is measured. This figure shows the spike frequency (as a function of time) in the form of a histogram. The adaptation of an auditory neuron can be summarised as follows.

- When a stimulus is applied suddenly, the spike rate response of the neuron rapidly increases. See region A on Figure 5.20.

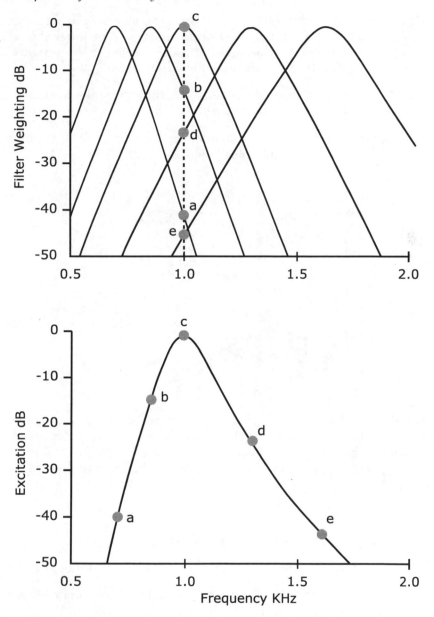

FIGURE 5.19: How the excitation pattern of a 1kHz pure tone can be generated through the auditory filter output as a function of the filter's centre frequency. The top figure shows five auditory filters (showing how the bandwidth of the filters increases with frequency) and the bottom shows how the excitation pattern can be obtained from the filter's output at 1KHz. Reprinted with permission. From: B Moore and BR Glasberg. Suggested formulae for calculating auditory-filter bandwidths and excitation patterns. *J Acoust Soc Am.*, 3(74):750-753, 1983. Copyright 1983, Acoustic Society of America [10].

- For a steady audio tone, the rate decreases exponentially to a steady state. See region B on Figure 5.20. At approximately 20ms after the tone is applied, a steady rate is achieved. This timescale can be recognised as correlating with the temporal masking effect shown in the next chapter.

- After the stimulus is suddenly removed, the tone spike rate reduces to below the spontaneous rate. See region C in Figure 5.20.

- It therefore can be concluded that an auditory neuron is often more responsive to changes than to steady inputs.

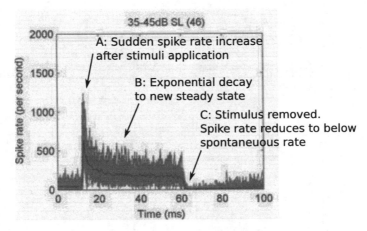

FIGURE 5.20: Adaptation of an auditory neuron. This figure shows a Post-Stimulus Time Histogram (PSTH) for an auditory neuron. Reprinted with permission. From Christian J Sumner and Alan R Palmer. Auditory nerve fibre responses in the ferret. *European Journal of Neuroscience*, 36(4):2428-2439, 2012 [17].

5.6.2　Frequency Response of the Auditory Nerves

Figure 5.22 shows a single tuning curve for an example auditory neuron.

- This graph was created by applying a 50ms tone burst every 100ms.

- The sound level is increased until the spike discharge increases by 1 spike per second.

- Repeat for all frequencies (on the x axis).

- Curves determined mainly by BM motion and the position of the hair cell that innervates the neuron.

5.6.3 Synchrony of the Auditory Nerves

Neuron spikes tend to occur in synchrony with applied stimuli, i.e., they are phase locked. However, phase locking does not exist above 5kHz and gradually diminishes above 1kHz.

5.6.4 Two-Tone Suppression

Two-tone suppression can be defined as a tone creating a steady state spike rate response having the spike rate decreased or suppressed by another tone [14] (illustrated in Figure 5.21).

- A short while after the tone starts, the response is strongly suppressed followed by a gradual increase to a new steady state.

- Upon removal of the new tone, the response increases suddenly to a peak before reducing to the previous steady state.

- Shaded areas in Figure 5.22 show the frequency of a tone able to suppress an existing tone.

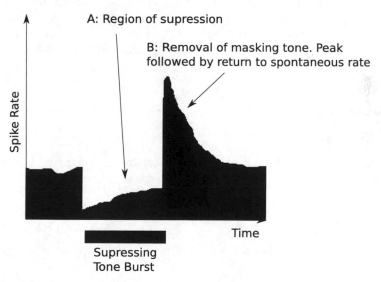

FIGURE 5.21: Two-tone suppression of an auditory neuron. This figure shows indicative results of a Post-Stimulus Time Histogram (PSTH).

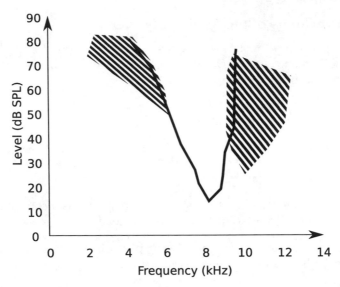

FIGURE 5.22: Tuning curve of a neuron and the regions of two-tone suppression. The main line shows a tuning curve of an auditory neuron with the characteristic frequency of 8000Hz. This figure also shows the shaded frequency regions of two-tone suppression. Reprinted with permission. From RM Arthur, RR Pfeiffer, and N Suga. Properties of "two-tone inhibition" in primary auditory neurones. *The Journal of Physiology*, 212(3):593-609 1971 [1].

5.7 Summary

- The physiology of the human hearing system is known as the periphery and comprises the components within the outer, middle and inner ear.

- Human frequency-based analysis is split into an overlapping set of auditory filters known as critical bands.

- These critical bands have a bandwidth that increases with central frequency.

- The shape of the auditory filters is obtained using psychoacoustic experiments such as those using notched filters.

- The auditory filters are approximately symmetric and can be described using models such as the RoEx filter.

- Although auditory filters are often modelled as symmetric, it has been found that the filters are skewed toward lower frequencies and the amount of skewness increases with the central frequency of the filter.

- Frequency based masking functions used in compression are the result of the response of a range of auditory filters, resulting in a masking function that has a wider skirt above the masking signal frequency and a narrower skirt below it.

- The critical bandwidths defined within the Bark scale are approximately equal to 50Hz below a central frequency of 100Hz but increase from that point approximately in 1/3 octaves. However, more accurately measured bandwidths using the ERB concept and notched noise experiments have shown that the bandwidths continue to decrease in size below 100Hz. Equation (5.14) is the most accurate model for critical bandwidths considered within this chapter.

5.8   Exercises

Exercise 5.1
Formulate the inverse equation of (5.2) and therefore obtain the centre frequencies that correspond to the following Bark bandwidths: {500,1000,1500}.

Exercise 5.2
Formulate the inverse equation of (5.3) and therefore obtain the frequencies that correspond to the following Bark values: {5,6,7}.

Exercise 5.3
Sketch the approximate masking function shape (excitation) of a masking tone at 0.8kHz using the same filters drawn on Figure 5.19.

Bibliography

[1] R.M. Arthur, R.R. Pfeiffer, and N. Suga. Properties of "two-tone inhibition" in primary auditory neurones. *The Journal of Physiology*, 212(3):593–609, 1971.

[2] D. Cunningham. *Cunningham's Textbook of Anatomy*. W. Wood, 1818.

[3] S. Young et al. The HTK Book. *Engineering Department, Cambridge University*, 2006.

[4] H. Fletcher. *Reviews of Modern Physics*, (12):47–61, 1940.

[5] B.R. Glasberg and B. Moore. Derivation of auditory filter shapes from notched-noise data. *Hearing Research*, 47(1):103–138, 1990.

[6] G.P. Schooneveldt and B.C.J. Moore. Comodulation masking release for various monaural and binaural combinations of the signal, on-frequency, and flanking bands. *J Acoust Soc Am.*, 1(85):262–272, 1987.

[7] H. Gray. *Gray's Anatomy*. 1858.

[8] H. Hermansky and N. Morgan. Rasta processing of speech. *IEEE Transactions on Speech and Audio Processing*, 2(4):578–589, 1994.

[9] P.O. Thompson J.C. Webster, P.H. Miller and E.W. Davenport. The masking and pitch shifts of pure tones near abrupt changes in a thermal noise spectrum. *J Acoust Soc Am*, pages 147–152, 1952.

[10] B. Moore and B. Glasberg. Suggested formulae for calculating auditory-filter bandwidths and excitation patterns. *The Journal of the Acoustical Society of America*, 74(3):750–753, 1983.

[11] B.C.J. Moore. Frequency analysis and masking. *Hearing*, pages 161–205, 1995.

[12] R.D. Patterson. Auditory filter shape. *J Acoust Soc Am*, pages 802–809, 1974.

[13] R.D. Patterson. Auditory filter shapes derived with noise stimuli. *J Acoust Soc Am*, pages 640–654, 1976.

[14] M. Sachs and N. Kiang. Two-tone inhibition in auditory-nerve fibers. *The Journal of the Acoustical Society of America*, 43(5):1120–1128, 1968.

[15] M.R. Schroeder, B.B. Atal, and J.L. Hall. Optimizing digital speech coders by exploiting masking properties of the human ear. *The Journal of the Acoustical Society of America*, 66(6):1647–1652, 1979.

[16] S.S. Stevens, J. Volkmann, and E.B. Newman. A scale for the measurement of the psychological magnitude pitch. *The Journal of the Acoustical Society of America*, 8(3):185–190, 1937.

[17] C. Sumner and A. Palmer. Auditory nerve fibre responses in the ferret. *European Journal of Neuroscience*, 36(4):2428–2439, 2012.

[18] H. Traunmuller. Analytical Expressions for the Tonotopic Sensory. *J. Acoust. Soc. Am. Scale*, 88:97–100, 1990.

[19] G. Von Békésy. *Experiments in Hearing*. McGraw-Hill series in Psychology. McGraw, 1960.

[20] S. Wang, A.Sekey, and A. Gersho. An objective measure for predicting subjective quality of speech coders. *IEEE Journal on Selected Areas in Communications*, 10(5):819–829, 1992.

[21] E. Zwicker. Subdivision the Audible Frequency range into Critical Bands. *The Journal of the Acoustical Society of America*, 33(2):248, 1961.

[22] E. Zwicker and E. Terhardt. Analytical expressions for critical-band rate and critical bandwidth as a function of frequency. *The Journal of the Acoustical Society of America*, 68(5):1523–1525, 1980.

6

Fundamentals of Psychoacoustics

CONTENTS

> We live in the digital age, and unfortunately it's degrading our music, not improving.
>
> Neil Young

Many different listening models based on the fundamentals of psychoacoustics are introduced within this chapter reflecting the many different aspects of the auditory system. Particular attention is given to the limitations of the human auditory system such as masking effects; such effects being able to be exploited for applications such as wideband audio coding.

6.1   Subjective Characteristics of Audio Signals

Three of the most common subjective characteristics of a sound are loudness, pitch and timbre [2]. These characteristics correspond to the measurable counterparts: sound intensity, fundamental frequency and overtone structure, respectively. Experiments show that a simple one-to-one relationship doesn't exist between the two sets.

Figure 6.1 shows a major direct relationship (signified by the double width arrows) between the two sets of characteristics. The dotted lines show the interdependence of all these terms (e.g., the subjective measure of loudness is directly related to sound intensity but can be also influenced by frequency and overtone structure).

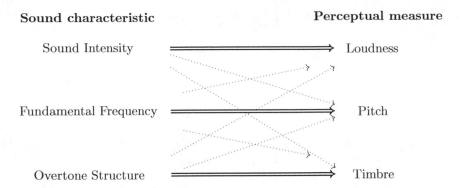

FIGURE 6.1: Relationship between sound characteristics and perceptual measures.

6.2 Loudness

Loudness is a subjective assessment of the magnitude of an auditory sensation and depends primarily on the amplitude of a sound.

Loudness is a psychological term that is universally understood to mean the perceptual magnitude of a sound. Specifically, measures of loudness are able to be ordered in magnitude giving rise to a relative scale of perceived magnitude. Common measures of loudness such as "very loud", "loud", "moderately loud", "moderately soft" and "very soft" can be represented by the musical notations *ff*, *f*, *mf*, *p* and *pp*. However, there is no universal agreement of how to define these terms and their notational counterpoints.[1] Table 6.1 shows a range of such measures with some suggestions of how they can be interpreted (together with suggested velocity values). However, not only is such a chart a long way from any formal definition, the associated perceptual loudness of any of these sounds depends on a multitude of contextual and subjective factors (such as sound duration and the age of the listener for instance).

---

[1]These musical dynamic notations (such as *mf* and *p*) do not signify an absolute loudness scale, but are utilised by composers to indicate intended relative-loudness within a piece. More recently they have become associated with defined note velocity values. For example, within Sibelius (a standard musical composition and transcription application), the velocities of each dynamic notation is strictly defined (as shown in table 6.1). Sometimes such tables include SPL loudness values in decibels. However, these values can only be nominally selected for reasons that will become clear below.

TABLE 6.1: Musical notation of score dynamics together with their interpretation, Italian equivalent and the default velocity value of the standard Sibelius notation application. [2]

ppp	Very Quiet	Pianississimo	20
pp	Somewhat Quiet	Pianissimo	39
p	Quiet	Piano	61
mp	Moderately Quiet	Mezzo-Piano	71
mf	Moderately Loud	Mezzo-Forte	81
f	Somewhat Loud	Forte	98
ff	Loud	Fortissimo	113
fff	Very Loud	Fortississimo	127

Loudness is also often confused with objective measures such as sound pressure, sound pressure level (SPL), sound intensity and sound power.

6.2.1 Sound Pressure Level, Sound Intensity, Phons and Sones

Sound pressure (also known as acoustic pressure) is defined as the local pressure deviation from average atmospheric pressure. The SI units of sound pressure are the same as pressure, i.e., the Pascal (Newton's per square metre: N/m^2).

Sound intensity (also known as acoustic intensity) is defined as the power passing through a unit area, normal to the direction of the sound wave. The SI units of sound intensity are the Watt per square metre (W/m^2).

As the power of a wave is proportional to the square of the amplitude, sound intensity is proportional to the square of the sound pressure:

$$I \propto P^2. \tag{6.1}$$

The most common measure of sound pressure as a sound characteristic is the Sound Pressure Level (SPL) and is defined as:

$$SPL = 10 \log_{10}\left(\frac{p^2}{p_0^2}\right) = 20 \log_{10}\left(\frac{p}{p_0}\right) \text{ dB,} \tag{6.2}$$

where p_0 is the reference pressure approximately equal to the threshold of hearing at 2kHz. It is defined to be $20\mu N/m^2$.

As sound intensity is proportional to the square of the pressure (see Equation (6.1)) the Sound Pressure Level can be defined as identical to the Sound Intensity Level (SIL). SIL is defined as:

$$SPL = SIL = 10 \log_{10}\left(\frac{I}{I_0}\right) \text{ dB,} \tag{6.3}$$

where I_0 is the reference sound intensity. It is defined to be $10^{-12} W/m^2$ and is the sound intensity of a wave with the reference sound pressure p_0.

6.2.2   A-weighting

A-weighting is a very common form of frequency compensation for the measurement of sound pressure levels. It is used to compensate for human perceived loudness, specifically to take into account the reduced sensitivity of the human ear to lower frequencies and to a lesser extent, very high frequencies. However, analysing the shape of the equal loudness curves (such as those shown in Figures 6.3 and 6.4) shows that their shapes vary significantly according to the loudness/intensity of the reference tone. Despite this, A-weighting is a single curve, functioning as a universal weighting scheme for all frequencies and intensities so that it can easily be utilised in many contexts.

A-weighting is defined as follows:

$$R_A(f) = \frac{12200^2 \cdot f^4}{(f^2 + 20.6^2)\sqrt{(f^2 + 107.7^2)}\,(f^2 + 737.9^2)(f^2 + 12200^2)}, \quad (6.4)$$

$$A(f) = 2.0 + 20\log_{10}(R_A(f)), \quad\quad\quad\quad\quad\quad\quad (6.5)$$

where f is the considered frequency and $A(f)$ is the A-weighting of frequency f. This curve is presented as a dashed line on Figure 6.4 (although it is more commonly presented in an inverted form). This inverted form is shown in Figure 6.2. This is directly from (6.5) and derived from the MATLAB code shown in Listing 6.1.

Listing 6.1: MATLAB code to create Figure 6.2 from (6.5)

```
1  f = 30:20000
2  RA= (12200^2 * f.^4)./((f.^2+20.6^2).*(((f.^2+107.7^2).*(f
      .^2+737.9.^2)).^0.5).* (f.^2+12200^2));
3  AF = 2.0+20 * log10(RA);
4  aw1 = plot(log2(f),AF);
```

A-weighting was developed shortly after the definition of the Fletcher–Munson equal loudness curves and was based on the 40 Phon curve of that work (shown as the nearest dashed line to the A-weighting curve in the same figure). Its use has been subsequently criticised as it has been claimed not to effectively compensate the low-frequencies at high intensities in addition to not properly representing the perception of noise rather than simple tones (as were used when defining A-weighting). Subsequent updates of A-weighting have included B, C and Z-weighting together with the ITU-R 468 scheme. These schemes have modified the equalisation of A-weighting to take these criticisms into account.

---

[2]www.sibelius.com

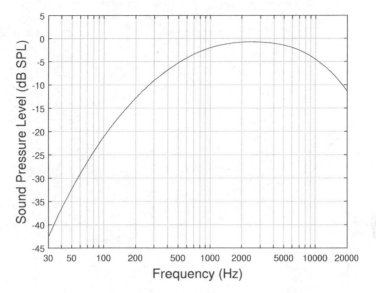

FIGURE 6.2: A-weighting curve created by MATLAB code in Listing 6.1 and Equation (6.5).

6.3 Equal Loudness Curves

Equal loudness curves are of key importance in the perceptual normalisation of sounds in the subsequent described audio applications (such as audio compression and speech recognition).

The unit of apparent loudness is the Phon. The Phon of a sound is defined as the SPL (in dB) of a reference tone at 1000Hz perceived as being equally loud. For example, 70 Phons means "as perceptually loud as a 1000Hz 70dB tone."

Many equal loudness curves exist [4]. Figure 6.4 shows a selection of equal loudness curves. All points on a single equal loudness curve share equal perceptual loudness across the frequency range of human hearing. Many differently defined equal loudness curves have been defined over the last hundred years including those originally defined by Fletcher and Munson in 1933 [3] and the most recently standardised curves defined by the ISO 226:2003 standard [1].

6.3.1 Fletcher–Munson Equal Loudness Curves

The Fletcher–Munson equal loudness curves have been taken for many years as the de-facto standard showing the variations of perceptual loudness with

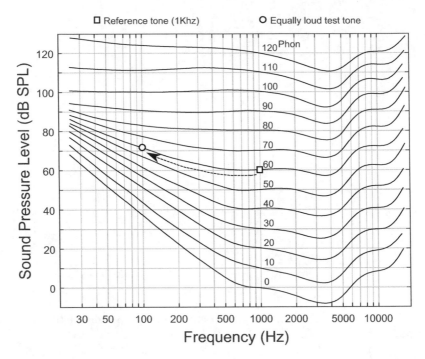

FIGURE 6.3: Fletcher–Munson equal loudness curves. Redrawn with permission from Fletcher, H., and Munson, W. J., "Loudness, its definition, measurement and calculation", *J. Acoust. Soc. Am.* 5: 82—108, 1933.

frequency and intensity. The procedure used to produce these curves is described concisely in the paper they produced in 1933 [3].

> The observers were seated in a sound-proof booth and were required only to listen and then operate a simple switch.....First, they heard the sound being tested, and immediately afterwards the reference tone, each for a period of one second. After a pause of one second this sequence was repeated, and then they were required to estimate whether the reference tone was louder or softer than the other sound and indicate their opinions by operating the switches. The level was changed and the procedure repeated. (Appendix A:104 [3])

Figure 6.3 shows the so-called Fletcher–Munson curves (defined in [3]). As described above, the curves are created by varying the intensity (sound pressure level) of the test tone until it matches the perceptual loudness of a reference tone at 1kHz. Reference sound pressure levels of {0, 10, 20, 30,. . . ,120} (shown on the figure at 1kHz) are therefore used to create each of the equal loudness curves.

A single comparison is shown in this figure where the 60dB test tone (at

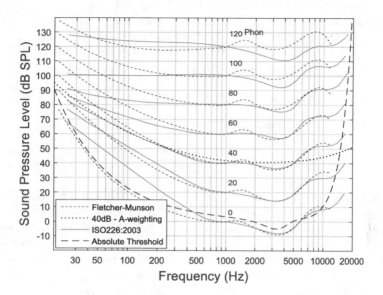

FIGURE 6.4: A variety of equal loudness curves. Data used from Figure 6.3 and ISO 226:2003 standard [1].

100Hz, shown as a circle) is matched to the reference tone (at 1kHz, shown as a square).

6.3.2 ISO 226

Since Fletcher and Munson's seminal 1933 paper "Loudness, Its Definition, Measurement and Calculation" [3] a multitude of equal loudness curves have been produced in the 20th century and beyond, many of these studies finding significant deviations from the previously assumed de-facto standard (especially in the lower frequencies). In an attempt to standardise equal loudness curves the international standard ISO 226 was defined in 1987. This was further refined in 2003 [1].

Figure 6.4 shows a comparison between the Fletcher–Munson curves and those loudness curves defined within the ISO-226/2003 standard. This clearly shows the significant differences between these curves at lower frequencies and higher volumes.

6.3.3 Sones and Phons

Two sounds with the same SPL measured in dBs will generally not be perceived as having the same loudness due to the issues of equal loudness across the frequency spectrum discussed above. The Phon is a frequency-equalised measure of loudness based on the concept of equal loudness curves. As with the equal loudness curves developed by Fletcher and Munson [3] shown in Figure 6.4 the reference tone is 1kHz and a sound is considered to have a loudness level of L_N in Phons if it is perceived as being as loud as a L_N dB tone at 1kHz. The implications of this definition are:

- Negative Phon values cannot be perceived.

- A Phon value of 0 defines the absolute threshold of hearing.

- A constant Phon value defines an equal loudness curve (as shown in Figures 6.3 and 6.4).

Although the Phon is a better definition of loudness as it is linear across the frequency scale, it is not directly proportional to perceptual loudness in terms of level. The Sone scale provides such a linear scale of perceptual loudness for all intensities. The Sone scale is defined to reflect the observation that firstly, an increase in Phon value of 10 is approximately heard as a doubling of perceived loudness and secondly, the most useful range of Phon values for musical and general listening is between 40 and 100 Phons.

The Sone scale is therefore defined as an exponent of 2 with the Phon value of 40 giving a Sone value of 1, 50 giving 2, 60 giving 4, etc. The relationship between Phons and Sones can, therefore, be defined by the following two equations (for Phon values above 40).

$$N = 2^{\frac{L_n - 40}{10}} \tag{6.6}$$

$$L_n = 40 + 10\log_2(N) \tag{6.7}$$

where L_n is the measurement of loudness made in Phons and N is the equivalent value in Sones. Table 6.2 and Equations (6.6) and (6.7) show the relationship between Phons and Sones. Although the dynamic notation symbols within this table are not strictly defined (as they are dynamic advice from the composer to the performer) this table shows how approximately they can be interpreted as Phon/Sone values.

6.4   Audio Masking

6.4.1   Absolute Threshold of Hearing (ATH)

For a given frequency there is an absolute threshold below which a noise or sound cannot be perceived (for the average person). This threshold is considered to be content independent. Although the ATH can be considered to be a specific equal loudness curve it has often been defined independently

TABLE 6.2: Relationship between Phons, Sones and Dynamic Notation

Phon	Sone	Dynamic Notation
40	1	*ppp*
50	2	*pp*
60	4	*p*
70	8	. . .
80	16	*f*
90	32	*ff*
100	64	*fff*

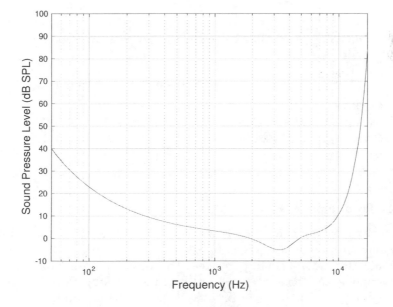

FIGURE 6.5: Absolute threshold of hearing: from Equation (6.8).

as it commonly forms a vital role in all forms of perceptual audio coding. The ATH is shown on the previously presented equal loudness curves shown in Figures 6.3 and 6.4 and can be approximated using the following formula defined by Terhardt [5] (where the output $ATH(f)$ is given in dB SPL):

$$ATH(f) = 3.64(f/1000)^{-0.8} - 6.5e^{-0.6(f/1000-3.3)^2} + 10^{-3}(f/1000)^4 \quad (6.8)$$

A single graph showing the ATH is shown in Figure 6.5.

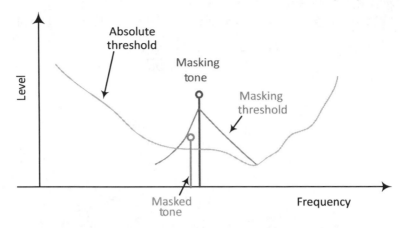

FIGURE 6.6: Auditory masking in the frequency domain.

Listing 6.2: Matlab code to create Figure 6.5 from (6.8)

```
1  f = 30:20000
2  at = (3.64*(f/1000).^(-0.8))-6.5*exp(-0.6*((f/1000-3.3).^2))
       +(10^(-3))*(f/1000).^4;
3  semilogx(f,at);
```

6.4.2   Frequency Masking

Frequency masking is most easily understood by the masking effect of one pure tone on another. If a quieter tone is within the masking threshold curve of a louder tone it will not be perceived. The masking threshold is dependent on frequency, i.e., the shape and width of the masking threshold curve of the louder tone is dependent on the frequency of the louder tone. Figure 6.6 illustrates frequency masking of one pure tone by another. This figure also shows the ATH. If the masked tone is underneath both the ATH or the masking threshold curve of the louder tone it will not be perceived.

6.4.3   Non-Simultaneous Masking

The perceptual masking of one signal by another is not limited to the time that the masking signal is present. There is a significant masking effect after the masking signal (post-masking) has been removed and a similar but smaller effect before (pre-masking). Although this may seem counter intuitive it can be easily illustrated by considering two synchronised metronomes. When played at the same tempo and within a small temporal offset window the two metronome sounds will be indistinguishable as being from two sources.

Figure 6.7 illustrates this type of non-simultaneous masking and shows

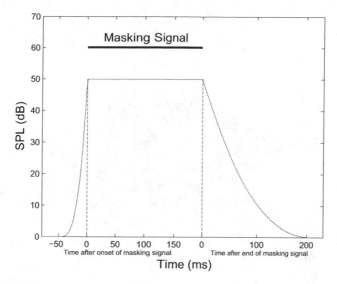

FIGURE 6.7: Non-simultaneous masking (temporal masking)

the shorter pre-masking and longer post-masking effects. The length of these effects have been extensively investigated and have been found to be between 20 and 50ms for pre-masking and 150–200ms for post-masking. This figure shows the shape of the masking as an exponential type. There is no universal agreement on the type of model to be used to define these "masking skirts". However, they are almost always illustrated in the same fashion.

Figure 6.8 shows the combination of the frequency and temporal masking effects. These graphs are an approximate combination of Figures 6.7 and 6.6 to form a 2D masking function dependent on both time and frequency. There have been reported specific effects of combining these forms of masking analysis but this has seldom been investigated or exploited.

6.5 Perception of Pitch

The perception of the pitch of a sound is a complex process within auditory perception. For pure tones, the perceived pitch corresponds to and varies with the frequency of the tone. For complex tones the perceived pitch is a result of a complex analysis of harmonic contents:

- Perception Place Cues: Peaks of excitation along the basilar membrane.

- Temporal Cues: The periodicity of each auditory nerve firing.

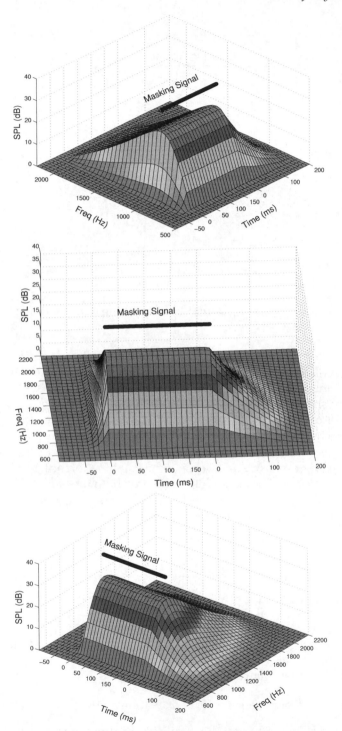

FIGURE 6.8: Temporal-frequency masking: Three angles from the same 2D temporal frequency masking function of an example masking tone.

- Variations in pitch create a sense of melody.

6.5.1   Perception of Pitch: Place Theory

The place theory of pitch perception is a theory that was originated by Helmholtz and assumes that the input sound stimulates different places on the basilar membrane and therefore activates specific auditory nerve fibres with characteristic frequency responses.

For complex tones with many frequency components, the many different maxima occur along the BM at positions corresponding to the frequencies of the components. The pitch is assumed to be the position on the BM with the maximum excitation.

However, for complex tones, the perceived pitch will not always correspond to the lowest frequency component present. For a complex harmonic sound, the pitch can remain perceptually constant even when its fundamental frequency has been removed.

6.5.2   Perception of Pitch: Temporal Theory

The temporal theory of pitch perception is based on the rate of neural firings for all the activated auditory neurons on the basilar membrane. The theory is based on phase locking of neural firings, the time intervals between the successive firings occurring at approximately integer multiples of the period of the waveform. Neural firing intervals are assumed to be measured by the superior olives or above in the auditory pathways. Periodicity can be extracted by autocorrelation.

However, phase locking does not occur above frequencies of 5kHz, therefore this theory does not explain the perception of complex tones above this frequency. Sounds produced by musical instruments, the human voice and most everyday sounds have fundamental frequencies less than 5kHz.

6.6   Psychoacoustics Summary

- The subjective sound characteristics {loudness, pitch, timbre} are the analogue of the machine measurable objective characteristics {sound intensity, fundamental frequency and overtone structure}.

- Frequency masking effects of the HAS

    - The ATH defines an absolute loudness of hearing. If a sound is quieter than this it cannot be perceived. Simple tones have an associated masking function. Underneath this masking function other tones cannot be perceived.

- Temporal masking effects of the HAS

  - Non-simultaneous masking effects can be measured for the HAS.

- The variation of perceptual loudness with frequency can be characterised by equal loudness curves (e.g., the Fletcher–Munson equal loudness curves).

- Perception of Pitch

  - Both the place of stimulation on the basilar membrane (place theory) and neural firing patterns (temporal theory) are important in determining the pitch of complex sounds.

  - The use of both forms of information for pitch perception is important and may depend on the harmonic content and type of sound.

  - Temporal information may dominate below 5kHz where phase locking occurs. Place coding may dominate for frequencies above 5kHz.

6.7   Exercises

Exercise 6.1
Write some MATLAB code to obtain the A-weighting of frequencies [100, 1000 and 10000]. Comment on this result.

Exercise 6.2
What frequency and equal loudness curves show the largest disparity between the Fletcher–Munson curves and the ISO226:2003 shown in Figure 6.4?

Exercise 6.3
Write some MATLAB code to convert between Phons (L_N) and Sones (N).

Exercise 6.4
Using the answer to the previous question find the values in Sones for the following Phon values 50, 60, 90, 130.

Exercise 6.5
Write some MATLAB code to convert between Sones (N) and Phons (L_N).

Exercise 6.6

Using the answer to the previous question find the values in Phons for the following Sone values: 1, 2, 3, 4.

Bibliography

[1] Acoustics-Normal Equal-Loudness-Level Contours. International Standard ISO 226: 2003. International Organization for Standardization, Geneva, Switzerland. 2003.

[2] H Fletcher. Loudness, pitch and the timbre of musical tones and their relation to the intensity, the frequency and the overtone structure. *The Journal of the Acoustical Society of America*, 6(2):59–69, 1934.

[3] H Fletcher and W Munson. Loudness, its definition, measurement and calculation. *Journal of the Acoustical Society of America*, 5(1924):82–108, 1933.

[4] D W Robinson and R S Dadson. A re-determination of the equal loudness relation for pure tones. *British Journal of Applied Physics*, 7:166–181, 1956.

[5] E Terhardt. Calculating virtual pitch. *Hearing Research*, 1(2):155–182, 1979.

7

Audio Compression

CONTENTS

> My goal is to try and rescue the
> art form that I've been practicing
> for the past 50 years.
>
> Neil Young

7.1 Introduction

7.1.1 Audio Compression: Context

The listening capabilities of human hearing are:

- Approximately 20kHz bandwidth perception

- 100dB dynamic range of intensity perception (sometimes a quoted maximum of 120dB).

For example, a Compact Disk (CD) uses 16 bits per sample. This bit-depth gives a possible dynamic range of $16 \times 6 \approx 96$dB SNR (given Equation (2.46))

TABLE 7.1: Quality, sampling rate, bits per sample, data-rate and frequency bands for five examples

Quality	Sampling Rate	Bits Per Sample	Data Rate	Frequency Band
Telephone	8kHz	8(Mono)	8.0kBps	0.2-3.4kHz
AM Radio	11.025kHz	8(Mono)	11.0kBps	0.1-5.5kHz
FM Radio	22.050kHz	16(Stereo)	88.2kBps	0.02-11kHz
CD	44.1kHz	16(Stereo)	176.4kBps	0.005-20kHz
DAT	48kHz	16(Stereo)	192.0kBps	0.005-20kHz

and is therefore adequate at covering the dynamic range of human hearing: 100dB.

7.1.2 Audio Compression: Motivation

Given the CD Sampling frequency of 44.1 kHz and sample size is 16 bit (2 bytes):

- Sound requires 1 sample per output, 2 samples for stereo

- 44,000[samples/sec] $\times$ 16[bits/sample] $\times$ 2[outputs] = 1,408,000 [bits/sec]

- Problem with standard digital format (CD)1408[kb/s] $\approx$ 176,400 [bytes/sec]

- Average song $\approx$ 32,000,000 bytes of space

Even in the present era of plentiful storage and bandwidth, this is still a significant amount of data when large numbers of songs/tracks are held together. To think of it another way, if this amount of data can be reduced six-fold and retain its perceptual quality (as most of the subsequently described algorithms can) then there can be six times as many tracks stored or six times as many DAB stations transmitted if the data is compressed using such types of audio compression. The economic motivation for audio compression is therefore obvious.

Table 7.1 shows a comparison of sampling rate, bits per sample, data-rate and frequency bands for five examples for a range of audio media. Many other higher sampling rates and bit-depths have also been used for uncompressed and compressed audio data (e.g., 96kHz and 24bit sampling).

Box 7.1.1: Why 44.1kHz?

Although the choice of approximately 40kHz is obviously justified as being twice the rate of the maximum human hearing frequency (20kHz). The actual choice of 44.1kHz may seem slightly arbitrary. It was however defined through the use of digital video tape machines for audio. As there were two common standards (NTSC and PAL), the choice of 44.1kHz was chosen as being easily encoded on both.

- NTSC (60 frames per second)

    - Encoding 3 samples on 245 active lines per field = $3 \times 60 \times 245 = 44100$ samples per second

- PAL (50 frames per second)

    - Encoding 3 samples on 294 active lines per field = $3 \times 50 \times 294 = 44100$ samples per second

7.2 Filterbanks for Audio Compression

The transformation of digital audio from "raw temporal forms" such as PCM or DPCM (defined in Section 11.2) in the time-frequency domain (and subsequent coding) is a universal feature within contemporary high performance wideband audio codecs. Such a time-frequency transformation results in a more sparse representation (for typical audio signals) and is therefore significantly easier to encode efficiently. This is because, firstly, due to the time-frequency domain being more statistically redundant than the time domain (in a large part due to typical sounds containing slowly varying tonal frequency components); secondly, the masking models described in the previous chapter can be accurately utilised within the time-frequency domain in order to most effectively exploit perceptual redundancies.

A common way to transform an audio signal into a time-frequency representation using a filterbank such as an M-band filterbank is illustrated in Figure 7.1. In order to avoid increasing the data-rate, the output of each filter in an M-band filterbank is subsampled by a factor of M resulting in a critically sampled/maximally decimated representation (i.e., the total number of subband samples is exactly the same as the number of input audio samples).

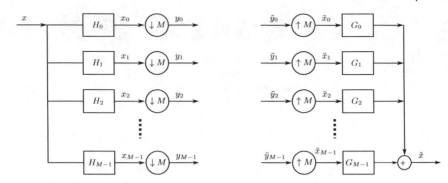

FIGURE 7.1: M-band multi-rate analysis and synthesis filterbank.

7.3 Perfect Reconstruction Filter-Banks

An important property of any filterbank is Perfect Reconstruction (PR, i.e., with reference to Figure 7.1, the input of the filterbank is identical to the output of the filterbank, i.e., $x = \tilde{x}$). In order to investigate this property we first examine solutions for two-band filter banks and then generalise to M-band filterbanks.

7.3.1 Two Channel Perfect Reconstruction Filterbanks*

Two channel digital filterbanks have often been used for audio coding; for example by Crochiere in 1976 [3] and then later in 1988 within the ITU standard G.722 [2]. Figure 7.2 shows the general structure of a typical two-channel multirate filterbank that is able to split the input signal into high- and low-frequency components.

FIGURE 7.2: 2-band filterbank.

The input signal x is filtered using a high and lowpass filter: H_0 and H_1, respectively. As this gives twice the data rate the filter outputs are subsampled by a factor of two (through the dropping of every other filter output as described in the z transform domain illustrated in section 2.7.3). This is possible without losing any information as the band-limited filter outputs will not be affected by the aliasing introduced by the subsampling. The synthesis filterbank is formed through the upsampling of $\tilde{y}_0$ and $\tilde{y}_1$ (through the interleaving of alternate zero values as described in the z transform domain in Equation (2.51)) and subsequent filtering of separately defined high and lowpass synthesis filters G_0 and G_1, respectively. Within an actual encoder, the analysis filter outputs y_0 and y_1 would be further processed and encoded (e.g., using quantisation and entropy coding) to create $\tilde{y}_0$ and $\tilde{y}_1$. However, we are interested at this stage in defining a perfect reconstruction system without any intermediate processing (i.e., $y_0 = \tilde{y}_0$ and $y_1 = \tilde{y}_1$) and therefore the challenge is to define the four filters H_0, H_1, G_0 and G_1 in order to provide perfect reconstruction.

Firstly, we consider the output $\tilde{x}$ in terms of the filter outputs $\tilde{y}_0$ and $\tilde{y}_1$:

$$\tilde{X}(z) = Y_0(z^2)G_0(z) + Y_1(z^2)G_1(z). \tag{7.1}$$

Additionally, we can generate expressions for Y_0 and Y_1 in terms of the input X and synthesis filters H_0 and H_1:

$$Y_0(z) = \tfrac{1}{2}\left(H_0(z^{1/2})X(z^{1/2}) + H_0(-z^{1/2})X(-z^{1/2})\right), \tag{7.2}$$
$$Y_1(z) = \tfrac{1}{2}\left(H_1(z^{1/2})X(z^{1/2}) + H_1(-z^{1/2})X(-z^{1/2})\right). \tag{7.3}$$

Finally, combining (7.1), (7.2) and (7.3) the following expression for perfect reconstruction for a two-band filterbank illustrated in Figure 7.2 is obtained. This condition must therefore be enforced during the designing of the analysis and synthesis filters (H_0, H_1, G_0 and G_1) to ensure perfect reconstruction:

$$\tilde{X}(z) = \frac{1}{2}X(z)\left[H_0(z)G_0(z) + H_1(z)G_1(z)\right],$$
$$+\frac{1}{2}X(-z)\left[H_0(-z)G_0(z) + H_1(-z)G_1(z)\right]. \tag{7.4}$$

Quadrature Mirror Filters (QMF)

In order to ensure that $\tilde{X}(z)$ equals $X(z)$ (i.e., perfect reconstruction considered within Equation (7.4)) the following two conditions must be met:

$$\begin{cases} H_0(-z)G_0(z) + H_1(-z)G_1(z) = 0, \\ H_0(z)G_0(z) + H_1(z)G_1(z) = 2. \end{cases} \tag{7.5}$$

The first condition ensures that the alias term is removed (the second line in (7.4)). It is ensured through defining the synthesis filters in terms of the analysis filters.

$$G_0(z) = -H_1(-z), \quad G_1(z) = H_0(-z) \tag{7.6}$$

In the time domain this implies that the synthesis highpass filter g_1 is identical to the analysis lowpass filter h_0 but with changed alternate signs of the filter taps. Also the synthesis lowpass filter g_0 is identical to the synthesis highpass filter with negative values together with changed alternate signs of the filter taps. This can be expressed as:

$$g_0[n] = -(-1)^n h_1[n], \qquad g_1[n] = (-1)^n h_0[n]. \tag{7.7}$$

To ensure the second PR condition in Equation (7.5) the QMF (Quadrature Mirror Filter) solution can be adopted. Starting with a definition of the lowpass synthesis filter H_0 the highpass synthesis filter is derived as follows:

$$H_1(z) = -H_0(-z) \quad \longleftrightarrow \quad h_1[n] = -(-1)^n h_0[n]. \tag{7.8}$$

This can be represented in the frequency domain as a reflection about the quadrature point $(\pi/2)$ in the frequency domain: [1]

$$H_1(\omega) = -H_0(\omega - \frac{\pi}{2}). \tag{7.9}$$

Therefore, once the lowpass analysis filter is defined (H_0), the three remaining filters can be derived (through the application of (7.7) and (7.8)) in order to generate a perfect reconstruction two-band system.

The ITU G722 standard is a key example of direct use of QMF filters. This standard is used for speech coding and was defined in 1988 with ADPCM coding being used on each QMF channel.

The filter H_0 is defined as an even-numbered linear phase (i.e., it is symmetric in the time domain) prototype filter. The remaining filters are defined from this filter and they are shown in the time and frequency domain in Figure 7.3.

Listing 7.1 shows the definition of the G722 filters from the H_0 filter, together with the two-channel multi-rate system using these filters within the structure shown in Figure 7.2. This code shows the close similarity between the input and the output signals. However, perfect reconstruction is not possible for FIR filters in such a structure (discounting the 2-tap Haar filter case which can be perfect reconstruction). [2] Perfect reconstruction is possible for such a two-channel structure, but a slight change is required to form the so-called Conjugate Quadrature Filter (CQF) solution.

---

[1] This is why the quadrature mirror filter is so-called as it mirrors across the quadrature point.

[2] Close to perfect reconstruction filters can be defined as in Listing 7.1: the linear phase FIR filters defined in G722.

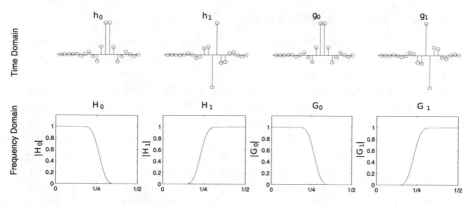

FIGURE 7.3: G722 QMF filters (as defined by Equations (7.7) and (7.8)).

Listing 7.1: Near perfect reconstruction two filter QMF filters

```
1   close all;
2   x = randn(256,1);
3   h0 = [12 -44 -44 212 48 -624 128 1448 -840 -3220 3804 15504];
4   h0 = [h0 wrev(h0)];
5   h0 = h0./sum(h0);
6
7   lh0 = length(h0);
8   k = 1:lh0;
9
10  h1=-((-1).^k).*h0;
11  h1 = (h1);
12  g0 = (h0);
13  g1 = -(h1);
14
15  y0 = downsample(conv(x,sqrt(2)*h0,'valid'),2);
16  y1 = downsample(conv(x,sqrt(2)*h1,'valid'),2);
17  g0output = conv(upsample(y0,2),sqrt(2)*g0,'valid');
18  g1output = conv(upsample(y1,2),sqrt(2)*g1,'valid');
19  reconSig = g0output +g1output;
20  plot(x(lh0:end),'k');
21  hold on;
22  plot(reconSig,'b+');
23  axis([0 100 -4 4]);
24  legend('Original','Reconstructed');
```

Conjugate Quadrature Filters (CQF)

Although the above QMF filter conditions should ensure perfect reconstruction it has been shown that this is only achieved in practice for FIR filters with filters having just two taps [10]. Therefore, only weak (in terms of frequency localisation) channel filters are available in the QMF case (e.g., Haar filters). It has been found that longer, more flexible FIR filters can satisfy the conditions and produce perfect reconstruction in practice. This is known as the Conjugate Quadrature Filter (CQF) solution. Similarly, for the case of QMF filters, the lowpass analysis filter is first defined (H_0) and the remaining filters are calculated from H_0 as follows. Firstly the analysis high-pass filter H_1 is the time reverse of h_0 modulated by $-(-1)^n$. The analysis filters g_0 and g_1 are just the time reverse of the analysis filters h_0 and h_1, respectively. This can be represented as:

$$h_1[n] = -(-1)^n h_0[N-1-n] \quad \longleftrightarrow \quad H_1(z) = z^{-(N-1)} H_0(-z^{-1}) \quad (7.10)$$
$$g_0[n] = h_0[N-1-n] \quad \longleftrightarrow \quad G_0(z) = z^{-(N-1)} H_0(z^{-1}) \quad (7.11)$$
$$g_1[n] = h_1[N-1-n] \quad \longleftrightarrow \quad G_1(z) = z^{-(N-1)} H_1(z^{-1}) \quad (7.12)$$

Listing 7.2 and Figure 7.4 shows an example CQF set of filters and codec structure. The figure shows how all the filters are derived from H_0 (e.g., time reversed and alternately negated).

CQF filters and structures form key components of iterated filterbanks such as Discrete Wavelet Transforms.

Listing 7.2: Perfect reconstruction two-filter CQF filters

```
1  close all;
2  x = randn(256,1);
3
4  load db10;
5  h0 = db10;
6
7  lh0 = length(h0);
8  k = 1:lh0;
9
10 h1=-((-1).^k).*h0;
11 h1 = wrev(h1);
12 g0 = wrev(h0);
13 g1 = wrev(h1);
14
15 y0 = downsample(conv(x,sqrt(2)*h0,'valid'),2);
16 y1 = downsample(conv(x,sqrt(2)*h1,'valid'),2);
17 g0output = conv(upsample(y0,2),sqrt(2)*g0,'valid');
18 g1output = conv(upsample(y1,2),sqrt(2)*g1,'valid');
19 reconSig = g0output +g1output;
```

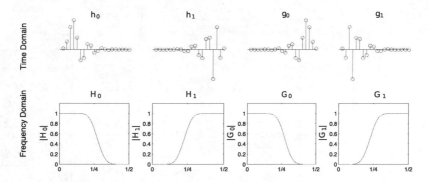

FIGURE 7.4: CQF filters (as defined by Equations (7.10), (7.11) and (7.12)).

```
20  plot(x(lh0:end),'k');
21  hold on;
22  plot(reconSig,'b+');
23  mean(abs(x(lh0:lh0+99)-reconSig(1:100)))
24  axis([0 40 -4 4]);
25  legend('Original','Reconstructed');
```

7.3.2   The Pseudo-QMF Filter-Bank (PQMF)

The QMF and CQF filter structures have been used to form two-band and iterated frequency decomposition of audio signals and can be structured in order to create perfect and near perfect reconstruction. In practice, however, biological plausibility would suggest the use of multiple (20–30 critical bands) subband decompositions. A Pseudo-QMF Filterbank (PQMF) structure is used within MP3 compression described below with 32 bands.

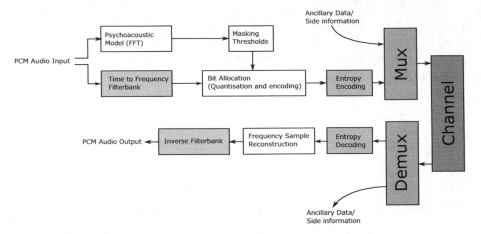

FIGURE 7.5: General perceptual based audio coder block diagram.

7.3.3   Wideband Audio Compression

There are numerous perceptually based wideband audio compression methods and standards. A common encoding and decoding structure for a wideband audio codec is illustrated in Figure 7.5. However, MPEG1 audio compression [5] is focussed upon, as it is one of the most common and well-understood compression standards and the principles of its underlying engineering methods are used almost universally within all other forms of wideband audio compression (e.g., AAC [6]). MPEG is an ISO standard. MPEG stands for Motion Pictures Expert Group. Although there are many MPEG standards, the most commonly used are MPEG1, MPEG2 and MPEG4. All three of these standards are used to compress both video and audio together.

- MPEG1. Designed to compress audio/visual data for static media (e.g., CDs, hard drives, etc.)

- MPEG2. Designed to compress audio/visual data for transmission (e.g., broadcast but also DVDs, etc.)

- MPEG4. Designed for higher compression performance and flexibility compared to MPEG1 and MPEG2)

7.3.4   MPEG1 Audio Compression

MPEG1 Audio Compression is a generic audio compression standard (it is not dependant on a source/filter vocal model). It has the following characteristics:

- Exploits perceptual redundancy

- Split into three layers. Each successive layer improves the compression performance at the cost of codec complexity

- Up to 6 to 1 compression ratio for perceptually unaffected compression

- It uses lossy compression. Lossy compression is a form of compression where there is always some form of loss when the original signal is compressed and uncompressed.

- Audio sampling rate can be 32, 44.1 or 48kHz

- The compressed bitstream can support one or two channels in one of four possible modes

    - a monophonic mode for a single audio channel,
    - a dual-monophonic mode for two independent audio channels (this is functionally identical to the stereo mode),
    - a stereo mode for stereo channels with a sharing of bits between the channels, but no joint-stereo coding, and
    - a joint-stereo mode that either takes advantage of the correlations between the stereo channels or the irrelevancy of the phase difference between channels, or both.

- It is separated into three layers. MPEG1 Layer 3 is what is generally known as MP3.

- The description of the standard is available from the ISO [5] and an excellent tutorial on the whole of the audio standard is given by Pan [7].

7.4   MPEG1 Audio Compression Layers

MPEG1 Audio Compression is separated into three separate layers. These layers target progressively increasing bit rates and have associated increased computational complexity.

Layer I

   - Computationally cheapest, targeted for bit rates > 128kbps
   - Used in Philips Digital Compact Cassette (DCC)

Layer II

   - Targeted for bit rate ≈ 128 kbps
   - Identical to **Musicam**

- It remains a dominant standard for audio broadcasting as part of the **DAB** digital radio and **DVB** digital television standards

Layer III

- Most complicated encoding/decoding, targeted for bit rates $\approx$ 64kbps, originally intended for streaming audio
- Predefined bit rates: 32, 40, 48, 56, 64, 80, 96, 112, 128, 160, 192, 224, 256 and 320 kbps

7.5 MPEG1 Audio Codec Structure

The high-level structure of the MPEG1 audio encoder has four elements common to all layers.

- Bank of polyphase (PQMF) filters: Frequency transforming filterbank

    - 32 subbands are created through a polyphase transformation to the frequency domain

- Psychoacoustic Model

    - Evaluates the perceptual relevance or irrelevance of each of the polyphase subbands

- Bit Allocator: Quantisation and coding

    - Codes the content of each polyphase subband according to the psychoacoustic model

- Frame Formatting

    - Creates the actual bit stream or output file

These four elements are illustrated to show how they fit together within the encoder in Figure 7.6.

7.5.1 Polyphase Filterbank

The polyphase filterbank is a PQMF filterbank that divides a signal into 32 equally spaced bands in the frequency domain. The filterbank is lossy even without quantisation. Each filter has a significant overlap of adjacent bands.

The polyphase filter inputs 32 new PCM samples for each computational cycle and outputs 32 values representing the output of 32 equally spaced filters (within the frequency domain). A FIFO buffer denoted as $x[n]$ (where

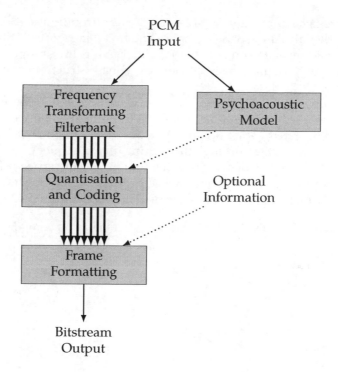

FIGURE 7.6: Wideband audio coding encoding structure: MPEG1 Audio.

n is the index 0:511) of 512 samples shifts in these new 32 PCM samples at each cycle.

The dot product of this FIFO buffer $x[n]$ with a windowing function $C[n]$ results in an intermediate buffer Z (also of length 512 samples).

$$Z[i] = C[i] \cdot x[i].$$

The partial calculation of $Y[k]$ is then calculated thus for $k = 0 \ldots 63$:

$$Y[k] = \sum_{j=0}^{7} Z[k + 64j].$$

The final 32 subband filter outputs $S[i]$ are then calculated thus for $i = 0 \ldots 31$:

$$S[i] = \sum_{k=0}^{63} Y[k] \cdot M[i][k].$$

The analysis matrix $M[i][k]$ is a collection of 64 partial sinusoids defined as:

$$M[i][k] = \cos\left[\frac{(2i + 1)(k - 16)\pi}{64}\right].$$

This entire calculation requires $512 + 32 \times 64 = 2560$ multiplies. All of the above stages of the calculation of S from x are illustrated in Figure 7.7. Window C is just a transformed version of a lowpass filter h. It is transformed in this way to enable the computationally efficient polyphase structure as illustrated in this figure.

This can be more easily understood in the equivalent (although not actually implemented and computationally more complex) method of obtaining S from x illustrated within Figure 7.8. This figure illustrates the fact that the 32 subbands are created through the shifting (in the frequency domain) of the lowpass filter h with sinusoids of different (and evenly spaced) frequencies. This all then results in the frequencies of all the 32 subbands shown in Figure 7.9. This figure is created using the code illustrated within Listing 7.3.

Listing 7.3: Polyphase filterbank implementation using MATLAB

```matlab
load C; %Load window C and original filter h

x = rand(512,1); %Input signal

Z = x.*C;
%Actual Polyphase Filter Implementation
j = 0:7;
for r = 1:64
    S(r) = sum(Z(j*64+r));
end

k = 0:63;
for i = 0:31
    M(i+1,:) = cos((1+2*i)*(k-16)*(pi/64));
end

for k = 1 : 32,
    y(k) = sum(M(k, :).*S);
end

%Direct Non-Polyphase Implementation
k = (0:511)';
for i = 0:31
    hi = h.*cos((1+2*i)*(k-16)*(pi/64));
    y(i+1) = sum(hi.*x);
end
```

The polyphase filterbank has the following shortcomings:

- Equal width filters do not correspond with critical band model of auditory system.

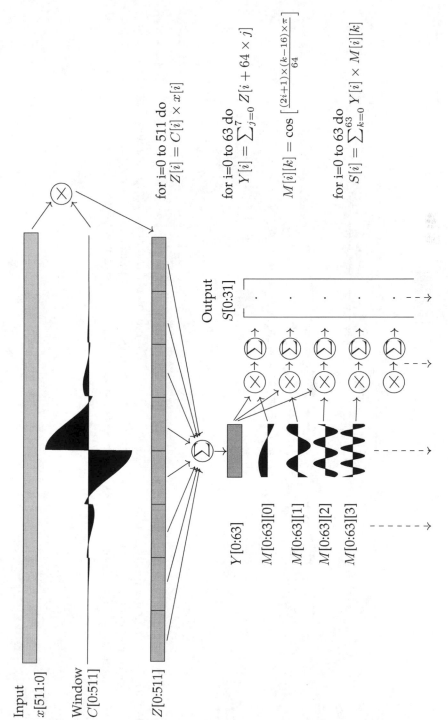

Input
$x[511:0]$

Window
$C[0:511]$

$Z[0:511]$

for i=0 to 511 do
$Z[i] = C[i] \times x[i]$

for i=0 to 63 do
$Y[i] = \sum_{j=0}^{7} Z[i + 64 \times j]$

$M[i][k] = \cos \left[\frac{(2i+1) \times (k-16) \times \pi}{64} \right]$

for i=0 to 63 do
$S[i] = \sum_{k=0}^{63} Y[i] \times M[i][k]$

Output
$S[0:31]$

$Y[0:63]$

$M[0:63][0]$
$M[0:63][1]$
$M[0:63][2]$
$M[0:63][3]$

FIGURE 7.7: 32-band polyphase filterbank.

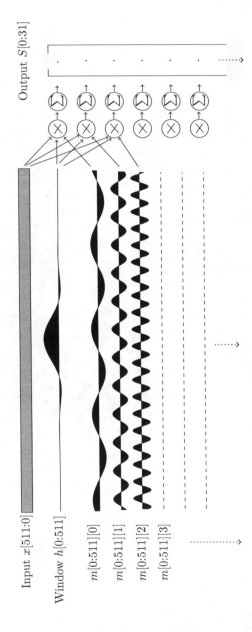

FIGURE 7.8: This figure illustrated how the 32-band polyphase filterbank is conceptually implemented: The FIFO input (x) is multiplied (samplewise) with the lowpass filter (h). This result is then multiplied (samplewise) with 32 different sinusoids (m) to create the 32 subband outputs (S). m is equivalent to M shown above, but extended to the length of the FIFO (512 samples).

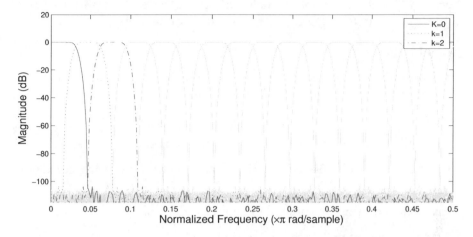

FIGURE 7.9: First three PQMF filters of the 32 filters within the polyphase filterbank of MPEG1: layers I, II and III.

- Filterbank and its inverse are *not* lossless (even without any further quantisation/compression).

- Frequency of critical bands overlap between subbands.

7.6 MPEG1 Audio Compression Psychoacoustic Models

7.6.1 Frequency Masking

Frequency masking as described in the previous chapter is used within the MPEG1 audio compression structure. As previously described, a single signal on the frequency plane will mask a neighbouring signal on the frequency plane. A masked signal must be louder than a determined "masking threshold" for it to be heard. Masked signals are therefore perceptually less important and therefore they are able to be more heavily quantised. A subband is quantised according to the "safe level for quantisation" obtained by the minimum masking threshold for that subband. The masking is greater for signal tones rather than signal noise. This is illustrated in Figure 7.12.

There are two separate psychoacoustic models implemented within MPEG1 audio compression. Each psychoacoustic model requires an independent time-to-frequency mapping to the polyphase filters at a finer frequency resolution than the polyphase filters. An FFT transform is used for both models implemented with a Hann window. Model 1 uses 512 samples. Model 2 uses two 1024 sample windows. Masking is handled significantly differently between "tones" and "noise" analysed within these FFT transforms since there is a greater masking effect of a tone rather than noise.

Model 1

- Tones are identified as local peaks. The remaining spectrum is "lumped" together per critical band into a noise signal at a representative frequency.

Model 2

- Tones and noise signals not specifically separated.

- A "tonality" index used to determine the likelihood of each spectral point being a tone.

- Calculated using two previous analysis windows.

- More robust than the method used in model 1.

For both models, spectral values are transformed into the perceptual domain. Spectral values are grouped together into values related to critical bandwidths. The masks generated from each of the detected tones (and residual noise) are then generated using the mask models shown in Figures 7.10 and 7.11. This, together with a mask modelling the ATH, is then integrated to give the overall masking.

The masking threshold is now calculated for each subband of the polyphase filterbank. The masking (or "spreading") function for Model 1 and 2 is illustrated in Figure 7.10 (for the variation of frequency) and Figure 7.11 (for the variation with intensity). These diagrams also show the previously popular masking function developed by Schroeder et al. [8].

Model 1

- A suboptimal system that considers all the masking thresholds within each subband and selects the minimum value.

- As the subbands are linearly distributed in the frequency domain this is an inaccurate method as the critical bands get wider as the frequency increases.

Model 2

- The above method is used if the subband is wider than the critical band.

- If the critical band is wider than the subband the average masking threshold is used.

The signal to mask ratio (SMR) is now calculated for each of the 32 subbands as:

$$SMR = \text{signal energy} \ / \ \text{masking threshold}. \tag{7.13}$$

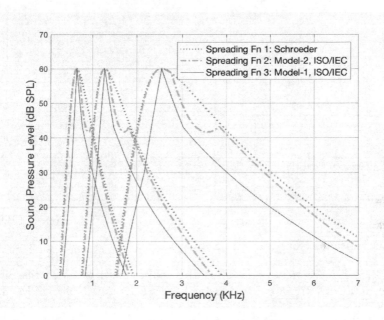

FIGURE 7.10: Frequency masking models: Variation of spreading function with frequency: MPEG1 Audio Model 1, 2 [5] and Schroeder et al. [8].

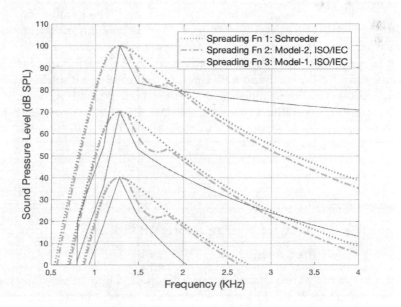

FIGURE 7.11: Frequency masking models: Variation of spreading function with intensity: MPEG1 Audio Model 1, 2 [5] and Schroeder et al. [8].

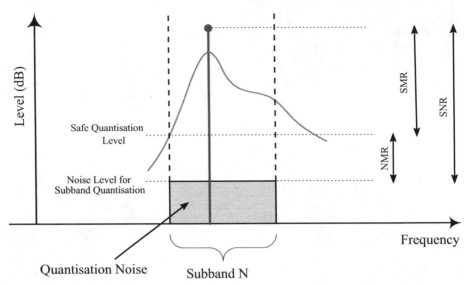

FIGURE 7.12: Frequency masking: The relationship between the Signal to Noise Ratio (SNR), Signal to Mask Ratio (SMR) and Noise to Mask Ratio (NMR).

The resulting SMR for each of the 32 coding bands are then passed to the coding module to allocate the bits to each subband. The relationship between SNR and SMR is illustrated in Figure 7.12. The generation of the final integrated masking function across all frequencies is illustrated in Figure 7.13. This figure is generated by the MATLAB code of Fabien Petitcolas [1] used to illustrate the use of the model 1 perceptual model within MPEG1 audio compression. This figure illustrates that the mask generation is done within a perceptually normalised frequency domain (i.e., the x axis) and the final mask being generated using the integration of all the individual (and ATH) masks (the dark dashed line).

7.7   MPEG1 Audio Compression: Bit Coding

7.7.1   Layer 1 Bit Coding

For layer 1 coding, the output bitstream is divided into frames. Each frame is the grouping together of 12 samples for each of the 32 subbands. This therefore results in a total of 384 samples per frame. According to the given quantisation calculated from the bit budget and utilised psychoacoustic model, each group

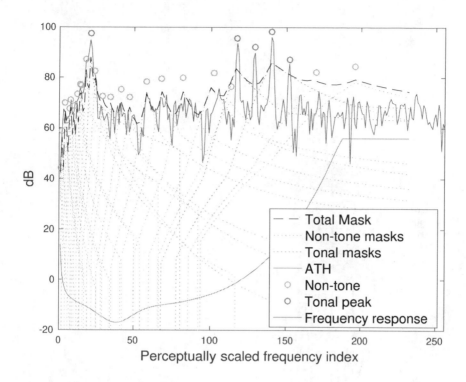

FIGURE 7.13: Frequency masking illustration showing the integrated masking function (dashed black line) generated from all of the tonal and non-tonal masking (spreading) functions (dotted lines). Generated from the MATLAB code of Fabien Peiticolas [1].

is encoded using from 0 to 15 bits per sample. Additionally, each group has a 6-bit scale factor.

7.7.2 Layer 2 Bit Coding

Layer 2 bit coding is very similar to layer 1 coding apart from

- The 12 sample groups per subband are grouped into three sets. There are therefore now 1152 samples per frame.

- There can be up to 3 scale factors per subband.

- The scale factors can be shared across subbands.

    - The information on whether to share scale factors across subbands is called Scale Factor Selection Information (SCFSI).
    - This can reduce the number of bits required.

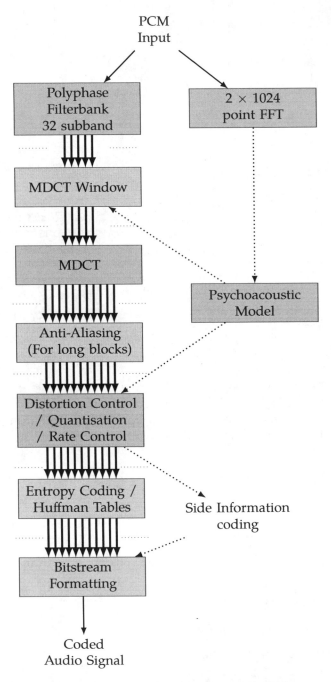

FIGURE 7.14: MPEG1 – Layer III (MP3) audio coding encoding structure.

7.7.3 Layer 3 Coding

Layer 3 coding is significantly more sophisticated than layers 1 or 2 (and is illustrated in Figure 7.14). Each of the 32 subbands are subdivided using the Modified Discrete Cosine Transform (MDCT) – a lossless temporal transform. In order to exploit the temporal masking effects of the HAS, Layer 3 coding uses an adaptive time window according to the temporal change of the audio content (see below).

Temporal Window Switching

The adaptive switching of window length and type is implemented in MP3 compression to reflect changes in the temporal range of approximate stationarity and therefore to suppress pre-echo and other temporal artefacts.

It is implemented by changing the analysis window length (L samples) from short durations (approximately 4ms) for transients to long durations (approximately 25ms) when the signal is more stationary. Therefore "long" (or normal) and "short" windows are defined together with "start" and "stop" windows for transitioning between the two. The decision as to when to change between the window lengths is made based on a threshold within the psychoacoustic model as detailed in the standard [5]. These four window types are illustrated in Figure 7.15 (generated by the code in Listing 7.4) and are defined as:

Normal Window: L=36, M=18

$$w_{normal}[n] = \sin\left(\frac{\pi}{36}(n + 0.5)\right) \quad \text{for } n = 0 \text{ to } 35. \tag{7.14}$$

Start Window: L=36, M=18

$$w_{start}[n] = \begin{cases} \sin\left(\frac{\pi}{36}(n+0.5)\right) & \text{, for } n = 0 \text{ to } 17 \\ 1 & \text{, for } n = 18 \text{ to } 23 \\ \sin\left(\frac{\pi}{12}(n - 18 + 0.5)\right) & \text{, for } n = 24 \text{ to } 29 \\ 0 & \text{, for } n = 30 \text{ to } 35 \end{cases} \tag{7.15}$$

Stop Window: L=36, M=18

$$w_{stop}[n] = \begin{cases} 0 & \text{, for } n = 0 \text{ to } 5 \\ \sin\left(\frac{\pi}{12}(n - 6 + 0.5)\right) & \text{, for } n = 6 \text{ to } 11 \\ 1 & \text{, for } n = 12 \text{ to } 17 \\ \sin\left(\frac{\pi}{36}(n + 0.5)\right) & \text{, for } n = 18 \text{ to } 35 \end{cases} \tag{7.16}$$

Short Window: L=12, M=6

$$w_{short}[n] = \sin\left(\frac{\pi}{12}(n + 0.5)\right) \quad \text{for } n = 0 \text{ to } 12. \tag{7.17}$$

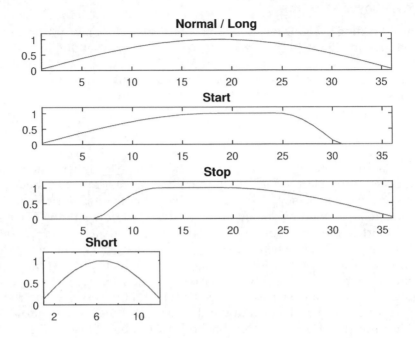

FIGURE 7.15: Four variable windows of the MDCT transform within the MP3 codec.

Listing 7.4: Code to define the four MDCT windows (normal, start, stop and short)

```
1  % block_type 0 (normal / long window)
2  ii = 0:35;
3  mdct_win0(ii+1) = sin(pi/36 * (ii + 0.5));
4  % block_type 1 (start window)
5  ii = 0:17;
6  mdct_win1(ii+1) = sin(pi/36 * (ii + 0.5));
7  ii = 18:23;
8  mdct_win1(ii+1) = 1;
9  ii = 24:29;
10 mdct_win1(ii+1) = sin(pi/12 * (ii - 18 + 0.5));
11 ii = 30:35;
12 mdct_win1(ii+1) = 0;
13
14 % block_type 3 (stop window)
15 ii = 0:5;
16 mdct_win2(ii+1) = 0;
17 ii = 6:11;
```

```
18  mdct_win2(ii+1) = sin(pi/12 * (ii - 6 + 0.5));
19  ii = 12:17;
20  mdct_win2(ii+1) = 1;
21  ii = 18:35;
22  mdct_win2(ii+1) = sin(pi/36 * (ii + 0.5));
23
24  % block_type 1 (short window)
25  ii = 0:11;
26  mdct_win3(ii+1) = sin(pi/12 * (ii + 0.5));
27
28
29  subplot(4,1,1); plot(mdct_win0);xlim([1 36]); ylim([0 1.2]);
        title('Normal / Long');
30  subplot(4,1,2); plot(mdct_win1);xlim([1 36]); ylim([0 1.2]);
        title('Start');
31  subplot(4,1,3); plot(mdct_win2);xlim([1 36]); ylim([0 1.2]);
        title('Stop');
32
33  subplot(4,1,4); plot(mdct_win3);xlim([1 12]); ylim([0 1.2]);
        title('Short');
34  a = get(gca,'Position');
35  set(gca,'Position',[a(1) a(2)*0.8 a(3)/3 a(4)])
```

MDCT Transform

The MDCT transform used within MP3 compression is an overlapped transform of length L samples but returning M coefficients. Each subsequent transform overlaps the previous one by 50% (M samples) and therefore achieves perfect reconstruction. The analysis filter for the MDCT transform is defined as (disregarding normalisation):

$$h_k[n] = w[n] \cos \left[\frac{(2n + M + 1)(2k + 1)\pi}{4M} \right] \quad \text{for } n = 0 \text{ to } 2M-1, \qquad (7.18)$$

where $L = 2M$ is the total length of the transform (36 samples for the long windows and 12 samples for the short and M is the length of the overlap between two transforms (18 samples for the long window and 6 samples for the short), n is the time index $(0, L)$ and $w[n]$ is the temporal window.

The synthesis filter is defined as the time reverse of the analysis filter (disregarding normalisation):

$$g_k[n] = h_k(2M - 1 - n). \qquad (7.19)$$

The short window consists of three small blocks windowed separately with the MDCT length defined by L and M adjusting the length of the MDCT transform defined in (7.18). The transition from normal to short window lengths and the reverse is defined to provide perfect reconstruction.

Alias Reduction Filter

A simple cross-subband alias reduction filter is defined with [5] to reduce the aliasing effect of the 32 polyphase QMF filter outputs.

7.7.4   Bit Allocation

Each subband is quantised using a number of bits determined by the Signal to Mask Ratio (SMR) calculated from the psychoacoustic model.
Layers 1 and 2

- Calculate the Mask to Noise Ratio

    - MNR = SNR - SMR (in dB)

    - SNR given by MPEG1 standard (as function of quantisation levels)

- Iterate the allocation of bits until no bits are left

    - Allocate bits to subband with lowest MNR.

    - Re-calculate MNR for subband and iterate.

Layer 3

- Entropy coding employed for each subband.

- Quantised output samples are reordered and entropy coded using Huffman tables.

- If Huffman encoding results in noise that is in excess of an allowed distortion for a subband, the encoder increases resolution on that subband.

- The whole process is iterated.

- Huffman coding:

    - A codebook is used to encode symbols output from the encoding system. More probable symbols are given codebook entries with fewer bits than less probable symbols, therefore, reducing the number of bits to represent the information.

7.8 Contemporary Wideband Audio Codecs

Although MP3 compression has existed for nearly three decades it is a very common form of audio compression. Other more modern wideband audio codecs include AAC and Opus. AAC (Advanced Audio Codec [6]) has been standardized as part of the MPEG2 and MPEG4 standards in the 1990s. It uses a very similar method to MP3 compression but improves on both the filterbanks and psychoacoustic models to provide a measurable improvement in rate-distortion performance. Opus [9] is a much more recent standard that defines very efficient audio compression over a wide range of bitrates for speech-based audio to wideband audio. Opus is covered in more detail in Chapter 11.

7.9 Lossless Audio Codecs

With regard to the quote at the start of this chapter from Neil Young, there has been a recent move to emphasise the advantages of lossless audio formats. It has been argued that compression artefacts in wideband audio compression systems have a subtle and undermining effect on the audio quality of the de-compressed audio (no matter what the level of compression). Lossless audio compression and decompression results in exactly the same original file. Obviously, formats such as Wav and AIFF are lossless (they are uncompressed PCM formats). Many lossless audio compression systems such as FLAC, [3] MPEG4 ALS [4] and MPEG4 SLS [11] have been standardised. Many of these systems are based on transforms such as LPC prediction with lossless en-coded residuals. Many new lossless applications are now gaining popularity due to rapid decrease in bandwidth and storage costs, e.g., Pono. [4]

7.10 Summary

- Perfect reconstruction two-band multirate filterbanks is possible using anti-aliasing and QMF conditions.

- CQF and QMF two-channel filterbanks firstly define the lowpass analysis filter H_0 and derive the remaining filters from it.

---

[3]https://xiph.org/flac/
[4]https://www.ponomusic.com/

- QMF and CQF filterbanks have been used for simple audio coding methods and standards [2, 3].

- MP3 compression is a commonly used and easy to understand wideband audio compression codec. It uses a combination of PQMF and MDCT filters and an FFT based psychoacoustic model.

- More modern compression codecs include AAC and Opus codecs.

- Lossless audio compression formats such as FLAC and MPEG4 ALS are gaining popularity.

7.11 Exercises

Exercise 7.1
Download and run the MATLAB code avaible for the illustration of MP3 coding [1]. Generate a graph similar to figure 7.13 for a sound file of your choice.

Exercise 7.2
Why is the MP3 filterbank implemented as shown in figure 7.7 rather than the equivalent method shown in figure 7.8?

Bibliography

[1] MPEG for Matlab. http://www.petitcolas.net/fabien/software/. Accessed: 2018-01-10.

[2] CCITT (ITU-T) Rec. G.722. 7 kHz audio coding within 64 kbit/s. *Blue Book, vol. III, fasc.*, III-4:269–341, 1988.

[3] R.E. Crochiere. Digital coding of speech in subbands. *Bell Sys. Tech. J.*, 55(8):1069–1085, 1976.

[4] T. Liebchen and Y.A. Reznik. Mpeg-4 als: An emerging standard for lossless audio coding. In *Data Compression Conference, 2004. Proceedings. DCC 2004*, pages 439–448. IEEE, 2004.

[5] MPEG-1 Audio, Layer III. Information technology – Coding of moving pictures and associated audio for digital storage media at up to about 1,5 mbit/s – Part 3: Audio. *ISO/IEC 11172-3:1993*, 1993.

[6] MPEG-2 Advanced Audio Coding, AAC. Information technology – Generic coding of moving pictures and associated audio information – Part 3: Audio. *ISO/IEC 13818-3:1998*, 1998.

[7] D. Pan. A tutorial on mpeg/audio compression. *IEEE Multimedia*, 2:60–74, Summer 1995.

[8] M.R. Schroeder, B.S. Atal, and J.L. Hall. Optimizing digital speech coders by exploiting masking properties of the human ear. *The Journal of the Acoustical Society of America*, 66(6):1647–1652, 1979.

[9] Opus Codec: http://opus-codec.org/: accessed 2018-01-01.

[10] P.P. Vaidyanathan. *Multirate Systems and Filter Banks*. Prentice-Hall, 1993.

[11] R. Yu, R. Geiger, S. Rahardja, J. Herre, X. Lin, and H. Huang. Mpeg-4 scalable to lossless audio coding.

8

Automatic Speech Recognition: ASR

CONTENTS

> Speech is an arrangement of notes
> that will never be played again.
>
> F. Scott Fitzgerald

8.1 Speech Recognition: History

Speech recognition has had a long history over the last 100 years. However, only recently has the dream of "talking to a computer" become an effective reality. One of the most notable and earliest speech recognisers was Radio Rex. There were no serious attempts at ASR for another 30 years when Bell Labs developed an isolated digit recognition system [2]. From that point on, a gradual improvement in speech recognition methods have been achieved with increases in vocabulary and robustness. Current speech recognition systems using Deep Networks have a recognition performance very close to that of a human.

The following lists some major milestones in ASR over the last 100 years.

1922: Radio Rex: Early toy, single-word recogniser.

1939: Voder and Vocoder: Speech synthesis, etc. The Voder was a speech synthesis system developed by Homer Dudley. It used levers and controls to manipulate the electronically produced sound to artificially synthesise the sound of speech.

1952: Bell Labs: Isolated digit recognition from a single speaker [2].

1957: 10 syllables of a single speaker: Olson and Belar (RCA labs) [6].

1950s: Speaker-Independant 10-vowel recognition: (MIT) [3].

1980s: Worlds of Wonder's Julie Doll Speech recognition and speech synthesis toy/doll.

1990s: Dragon Dictate

2008s: Google Voice Search

2010s: Siri, Cortana, Alexa, Google Assistant

A comprehensive list and description of speech recognition systems is given by Junqua and Haton [5].

8.1.1   Radio Rex

Radio Rex was a speech recognition toy produced in the United States in 1922 by Elmwood Button Co. It was comprised of a dog made of celluloid attached to an iron base. It was kept inside a wooden kennel by an electromagnet against the force of a spring. The current to the electromagnet flowed through a loosely touching connector. This physical connection was sensitive to approximately 500 cps acoustic energy. This was approximately the energy of an average speaker saying the word "Rex" close to the toy. Therefore, when "Rex" was said, the connection was broken, the electromagnet turned off and the spring pushed the dog out the kennel.

8.1.2   Speech Recognition: History: 1952 Bell Labs Digits

The isolated single digit recogniser developed by Bell Labs was the first successful speech recognition system on a multiple vocabulary [2]. It approximates energy in formants (vocal tract resonances) for each spoken digit 0,1,2,3,4,5,6,7,8 and 9. Although rudimentary, it already integrated some robust ideas (insensitive to amplitude, timing variation, etc.). The main weaknesses in the system were found to be technological (the robustness of the physical circuits and components). A circuit diagram of the developed system (taken from [2]) is shown in Figure 8.1.

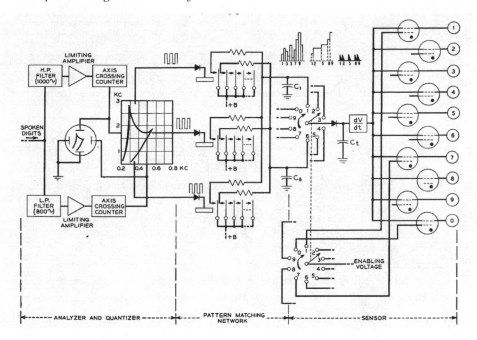

FIGURE 8.1: A circuit diagram of the digit recogniser developed at Bell Labs in 1952. The input signal is shown on the left and separated into high and low frequencies using a high and a lowpass filter. Lights labelled from 0 to 9 are illuminated based on the recognised digit (shown on the right). Reproduced with permission. From: KH Davis, R Biddulph, and Stephen Balashek. Automatic recognition of spoken digits. *The Journal of the Acoustical Society of America*, 24(6):637–642, 1952. Copyright 1952, Acoustic Society of America [2].

8.1.3 Advances 1960s–Present Day

The following technological advances have been key in developing the current state of the art speech recognition systems that are able to attain near human recognition performance. These methods are listed from the oldest to the most recent.

- Spectral analysis – FFT

- Cepstrum

- Dynamic time warp

- Hidden Markov Models (HMMs)

- Language models

- Integration with Deep Neural Network Methods

8.2   ASR-Problem Formulation

Automatic speech recognition poses some extremely difficult engineering problems. The problems can be categorised as follows (in increasing levels of difficulty):

- Single word recognition systems (e.g., the single digit recogniser by Bell Labs)

- Single speaker ASR

- Small dictionary systems

- Unconstrained large dictionary recognition tasks

Compact and invariant features are required to base recognition upon. Frequency representations need to be invariant to pitch changes, etc. An ASR system must be able to take into account of the following:

- Timing variation

    - Timing variation changes between speakers
    - Timing variation changes between the same speaker in different environments
    - Formal/casual speech

- Loud/quiet speech

- Between speaker variability

    – Accents
    – Gender
    – Vocal mannerisms

- Contextual effects

    – "How do you do?"
    – "I scream for ice cream."

Finally, each language may have special features that will need to be taken into account, e.g., the click consonants of the Khoisan African languages.

8.3 ASR Structure

8.3.1 Linguistic Categories for Speech Recognition

Feature vectors can be extracted from a known speech signal and used as a representation of linguistic categories (or sequence of categories) in the signal. The basic modelling units for speech recognition are based on these linguistic categories. Recognition is made possible through training and testing using the association of these linguistic categories with extracted feature vectors.

8.3.2 Basic Units for ASR

Initially, single words appear to be a natural unit for speech recognition systems. Using entire words rather than using individual subwords (phones) allows the use of the context of the phones to be covered in the model. In single small vocabulary problems, this is a good design as there are many instances for each word. However, many identical sub-word elements appear in more than one word. Word models ignore this commonality and for large dictionaries can be extremely redundant for effective recognition. Word units are therefore inappropriate in large-vocabulary systems. However, some work has been done on the recognition of phrases.

8.4 Phones

Phones are the phonetic sub-word units that are the most common basic units for automatic speech recognition systems. Phones are preferred as the basic unit for general ASR over words and phrases. However, word models can be constructed using concatenated phone models. Phone context can be modelled by context-dependent phone models. For example:

- English word cat is spelt in phones as [kæt].

8.4.1 Phones versus Phonemes versus Graphemes

Phonemes and Phones are often confused in speech recognition literature. Phonemes are abstract subword structures that distinguish meaning within a language. They are usually represented between slashes, e.g., /p/.

Conversely, a phone is an actual instance of a phoneme, i.e., how it is actually pronounced (i.e., they are only ever spoken or pronounced in one defined way). They are usually represented between square brackets, e.g.,

[p]. A number of phones that can be used to pronounce a phoneme are called "allophones". A basic audio to phone recognition auditory model is the very basic element of a speech recognition system with the conversion from phone to phoneme depending on accent, dialect and context. Description of auditory models below refers to phone recognition rather than phoneme recognition in the first instance.

A grapheme is a number of letters or a single letter that represents a phoneme. An alternative definition is to say that a grapheme is a number of letters or a single letter that spells a sound in a word. Graphemes have been more recently used for state of the art ASR methods.

8.5 Phonetic Alphabets

The most common phonetic alphabet is the International Phonetic Alphabet (IPA).

- 75 consonants

- 25 vowels

- Each phone is a defined character in square brackets, e.g., $[p],[t]$

- Most common phonetic alphabet reference

- Covers most languages

- Figure 8.2 shows a subset of this alphabet; just showing the vowels.

- A full chart of the IPA alphabet is available for free from the IPA. https://www.internationalphoneticassociation.org/

An alternative TIMIT phoneset contains

- 61 phones

- English specific

- Each phone defined as ASCII symbol (machine readable)

- Associated with the TIMIT ASR ground truth dataset

VOWELS

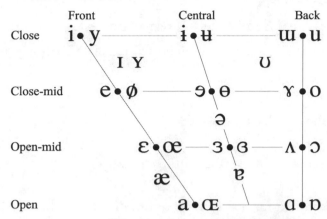

Where symbols appear in pairs, the one
to the right represents a rounded vowel.

FIGURE 8.2: The vowels of the International Phonetic Alphabet (IPA):
http://www.internationalphoneticassociation.org/content/ipa-chart
Copyright © 2015 International Phonetic Association, licensed under CC BY
3.0 https://creativecommons.org/licenses/by/3.0/.

8.6 Deterministic Sequence Recognition

For deterministic sequence recognition commonly, feature representations are
extracted every 10ms over a window of 20 or 30ms. Features are derived from
PLP or MFCC or other similar types (see Chapter 9). Deterministic sequence
recognition is based on word-based template matching as illustrated in Figure
8.3.

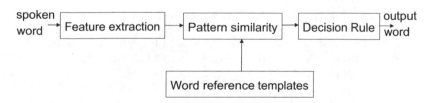

FIGURE 8.3: Deterministic word-based template matching.

- Can be used for both isolated and connected word recognition

- Deterministic: Supplanted by statistical methods

- Popular in late 1970s to mid 1980s

8.7 Statistical Sequence Recognition

Statistical sequence recognition can be defined mathematically as follows:

- W is a sequence of words (or phones): $w_1, w_2, \ldots, w_N$

- W^* is the most likely sequence

- X is a sequence of acoustic features: $x_1, x_2, \ldots, x_T$

    - Feature extraction is discussed in detail in section 2.9.1 and in the next chapter (for ASR).

- Θ is a set of model parameters.

The most likely sequence W^* is obtained using Maximum A-Posteriori (MAP) Bayesian system (given Bayes' rule (2.65)):

$$W^* = \operatorname*{argmax}_{W} P(W|X, \Theta)$$

$$W^* = \operatorname*{argmax}_{W} \frac{P(X|W, \Theta)P(W|\Theta)}{P(X)} \quad \text{Bayes' rule}$$

$$W^* = \operatorname*{argmax}_{W} P(X|W, \Theta)P(W|\Theta) \tag{8.1}$$

$P(X)$ is constant (known as the "evidence" term as given by the Bayes' rule).

8.8 Language and Auditory Models

Equation (8.1) defines how language and auditory models can be combined. This type of probabilistic model is known as a Maximum A-Posteriori (MAP) Bayesian system:

$$W^* = \operatorname*{argmax}_{W} \underbrace{P(X|W, \Theta)}_{\text{Auditory model}} \underbrace{P(W|\Theta)}_{\text{Language model}}$$

The output of the system is therefore not only dependent on the likelihood obtained from the auditory model $P(X|W, \Theta)$ but also on the probability (or prior) of each word or phone occurring in the context given by a language model $P(W|\Theta)$. Language models integrate a lexicon defining phone order within defined words and a grammar model for likely word orders. The auditory model/language model system was developed by Jelinek in 1976 [4].

8.9 Speech Recognition Datasets

The generation and evaluation of an ASR system require an accurately labelled speech dataset. There are several commonly used labelled speech recognition datasets (these examples are all English). These include:

TIMIT: The defacto standard small scale speech recognition database. 6300 sentences, 10 sentences spoken by each of 630 speakers with eight dialects.[1]

CMU AN4: Small speech database of single words: 948 training and 130 test utterances: 16 kHz, 16-bit linear sampling. [2]

Common Voice: A large Mozilla labelled speech dataset that is currently over 200 hours long. [3]

Switchboard Hub5′00: 40 English telephone conversations (used by [1]). [4]

8.10 Summary

- ASR has a history of over 100 years.

- The basic units of ASR include phones and phonemes.

- Language models are used to develop contextual ASR systems.

- Many speech recognition datasets are available for speech recognition evaluation.

---

[1]https://catalog.ldc.upenn.edu/ldc93s1
[2]http://www.speech.cs.cmu.edu/databases/an4/
[3]https://voice.mozilla.org/
[4]https://catalog.ldc.upenn.edu/LDC2002T43

Bibliography

[1] A. Hannun, et al. Deep speech: Scaling up end-to-end speech recognition. arxiv preprint arxiv:1412.5567. 2014.

[2] KH Davis, R Biddulph, and S. Balashek. Automatic recognition of spoken digits. *The Journal of the Acoustical Society of America*, 24(6):637–642, 1952.

[3] J.W. Forgie and C.D. Forgie. Results obtained from a vowel recognition computer program. *The Journal of the Acoustical Society of America*, 31(11):1480–1489, 1959.

[4] F. Jelinek. Continuous speech recognition by statistical methods. *Proceedings of the IEEE*, 64(4):532–556, 1976.

[5] J-C Junqua and J-P Haton. *Robustness in Automatic Speech Recognition: Fundamentals and Applications*, volume 341. Springer Science & Business Media, 2012.

[6] H Olson and H Belar. Phonetic typewriter. *IRE Transactions on Audio*, 5(4):90–95, 1957.

9

Audio Features for Automatic Speech Recognition and Audio Analysis

CONTENTS

> Every time I fire a linguist, the
> performance of our speech
> recognition system goes up.
>
> Fred Jelinek

9.1 Speech Features: Introduction

The input to any speech recognition system is the raw sampled recorded audio data. It is not practical or desirable to try to implement a recognition system using the raw input sampled audio data due to an extremely large amount of data redundancy in typical speech signals and a lack of any form of speaker/environmental invariance in the data.

A lower dimensional representation of the input signal is typically necessary in order to reduce the input redundancy and generate representative vectors that are invariant to environmental and speaker variability. Such compact representations should highlight and retain a maximum amount of speech distinguishing information and are known as features (or feature vectors, see Section 2.9.1).

Furthermore, the extraction of features must be achieved at a temporal resolution where the frequency characteristics of the signal can be considered to be stationary. Features are usually extracted for every "frame" of the input audio file. As is common in all the described audio analysis within this book "frames" are typically of the order 10–30ms.

Feature vector extraction is a universal pre-processing stage of any supervised and unsupervised machine learning algorithm (as described in Chapter 2). Supervised machine learning systems (such as speech recognition) require such representations extracted from a set of training examples and then matched to the same features extracted in the testing stage for actual recognition using a suitable classifier.

From the source-filter theory of speech production, it can be seen that the semantic content of speech is contained within the filter part of the signal (i.e., the frame's "spectral envelope"). This is due to the fact that the source is the vibration caused either by the vibrating glottis or the excitation caused by air passing through a constricted airway (such as the lips or between the tongue and the upper mouth). The source therefore obviously does not contain the semantic content of the speech signal as it can vary due to context and from speaker to speaker. Features that characterise speech content therefore usually target a representation of the filter.

9.2   Tools for Speech Analysis

9.2.1   Wideband versus Narrowband Filterbanks Spectral Analysis for Speech Audio

A very common and conventional way to analyse and extract features from speech-based audio is to use a filterbank across the relevant frequency range for speech. The main issue in using such a filter bank is the frequency resolution. Conventionally (for speech analysis) frequency resolution is separated into either "narrowband" or "wideband" categories. Figure 9.1 shows the issues in the use of narrowband and wideband filterbanks. In this case, a wideband analysis is required as indicated by the bottom two graphs in Figure 9.1. Wideband analysis aggregates the vertical source-excitation harmonic lines (see top graph in the figure) to produce the formant structure of the speech signal (i.e., the spectral envelopes). Conversely, narrowband analysis uses a bank of filters that are of a similar width to the excitation harmonic width (the distance between each harmonic) as illustrated in the same figure. In this case, the harmonic-source-excitation structure is revealed by the narrowband analysis (i.e., very narrow horizontal lines in a spectrogram).

Below, when using FFT transformations as a filterbank for the analysis of speech, guidelines are given as to what constitutes a narrowband and a wideband analysis, i.e., what frequency resolution extracts the source (excitation) and filter (spectral envelope), respectively.

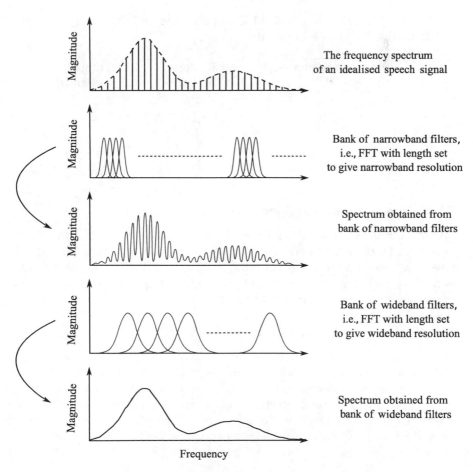

FIGURE 9.1: Narrowband versus wideband filterbanks for speech analysis.

TABLE 9.1: FFT lengths to give 30Hz (narrowband) or 200Hz (wideband) FFT bandwidths

Sampling Frequency (F_s)	FFT Length (Wideband)	FFT Length (Narrowband)
48000	1600	240
44100	1470	220
22050	735	110
11025	368	55
8192	273	41

TABLE 9.2: Closest power of 2, FFT lengths to give 30Hz (narrowband) or 200Hz (wideband) FFT bandwidths

Sampling Frequency (F_s)	FFT Length (Wideband)	FFT Length (Narrowband)
48000	2048	256
44100	1024	256
22050	512	128
11025	512	64
8192	256	32

9.2.2 Spectrograms for Speech Analysis

The most critical parameter of a spectrogram (or related Short-Time Fourier Transform) is the length of the FFT (mostly equivalent to the window length). Given that the window length and the length of the FFT are equal (i.e., there is no zero padding or truncation) this single parameter can determine the frequency resolution of the FFT filter bank.

Narrowband Spectrogram for Speech Analysis

Narrowband analysis of a speech signal is designed to highlight and visualise the harmonics of the speech source. These harmonics can be seen as narrow horizontal lines on a spectrogram with the lines near formant frequencies being highlighted. In order to effectively visualise these horizontal "source" based harmonic lines, the width of each FFT band within a spectrogram should be of the order of 30–45Hz, (corresponding to spectral snapshots of 20–30ms).

Wideband Spectrogram for Speech Analysis

Wideband analysis of a speech signal is designed to highlight and visualise the formants of the speech source. Formants can be seen as horizontal bands on a spectrogram (but much wider than the thin source harmonic lines). In order to effectively visualise these horizontal formants, the width of each FFT band within a spectrogram should be of the order of 200Hz, (corresponding to a spectral snapshot of 5ms). An analysis bandwidth of 200Hz is generally perfectly acceptable for most male voices. However, for cases where the first formant (F_1) of the voice is very high (for a high pitched male, woman or child), then 250Hz or even 350Hz may be preferable as an analysis bandwidth.

Table 9.1 shows some guidelines for the FFT length for creating narrowband/wideband spectrograms for different choices of sampling frequency (of the input signal). These lengths are approximate and in actual implementations may be constrained to be powers of 2 (i.e., 2^N where $N \in \mathbb{N}$). Table 9.2 shows the closest power of 2 FFT length to the exact lengths given in table 9.1. Figures 9.2, 9.3 and 9.4 show narrowband and wideband spectrograms of spoken words.

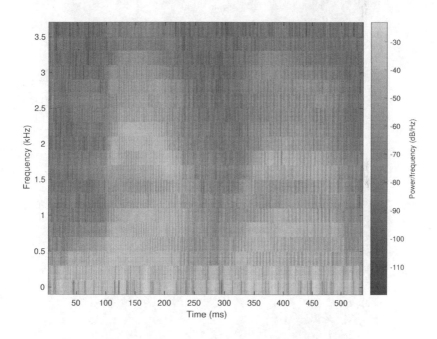

FIGURE 9.2: Spectrogram of Matlab word (wideband).

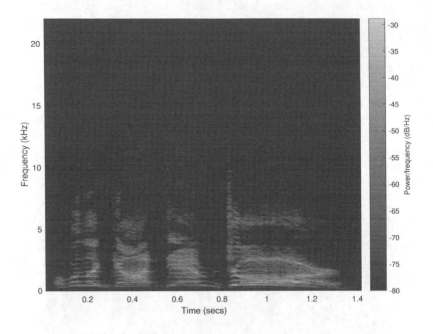

FIGURE 9.3: Spectrogram of Door word (narrowband).

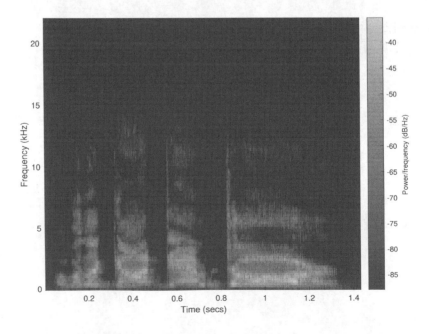

FIGURE 9.4: Spectrogram of Door word (wideband).

FFT Frequency Resolution

Given an FFT length (with no zero padding) N and a sampling frequency of F_s. The frequency resolution F_Δ of an FFT (the bandwidth of each FFT frequency bin) is given by

$$F_\Delta = \frac{F_s}{N}$$

For example if $F_s = 44100$Hz and N = 1024 the $F_\Delta \approx 43$Hz.

9.3 Cepstrum Analysis

The direct use of Fourier Transforms/Spectrograms/Short Term Fourier Transforms for generating speech features is problematic; it does not effectively separate the source and filter components of the speech signal.

The Cepstrum was developed by Oppenheim in the 1960s to separate the source and filter components of speech using spectral methods [5]. The source-filter theory of speech production states that the signal $s[t]$ is the convolution of the source $g[t]$ and filter $h[t]$ (assuming a discrete representation where g and h are indexed by time index t):

$$s[t] = g[t] * h[t]. \tag{9.1}$$

Where $*$ is the convolution operator. Given the convolution theorem of the Fourier transform and the fact that S, G and H are the Fourier transforms of s, g and h, respectively, leads to the result that the Fourier transform of the signal is equal to the multiplication of the Fourier transform of the source with the Fourier transform of the filter:

$$S(\omega) = G(\omega) \times H(\omega) \tag{9.2}$$

This goes some way to make the separation more tractable. However, in order to improve this situation we can take the log of both sides of (9.2) rendering the product into an addition:

$$\log(S(\omega)) = \log(G(\omega)) + \log(H(\omega)). \tag{9.3}$$

The key insight to understanding the Cepstrum transform is that the source signal will have a log frequency spectrum that varies more over the frequency scale than the filter. In order to separate the high varying and low varying components in the frequency domain a subsequent frequency analysis is performed and the result can be simply separated by truncation/thresholding. This further frequency transform is usually implemented using an inverse frequency transform which effectively does the same thing but has the positive effect of keeping the units more logical. Using the FFT as the transform the Cepstrum can be defined as follows:

$$\text{IFFT}(\log |\text{FFT}(s[t])|). \tag{9.4}$$

Figures 9.5, 9.6 and 9.7 illustrate the different stages of the Cepstrum transform. An example of these stages are implemented in the MATLAB code shown in Listings 9.1 and 9.2 generating the following figures. The truncation code and figure show that retaining a small number of Cepstrum coefficients can closely model the spectral envelope of the input speech signal (with the larger number of coefficients leading to a closer fitting to the spectral envelope).

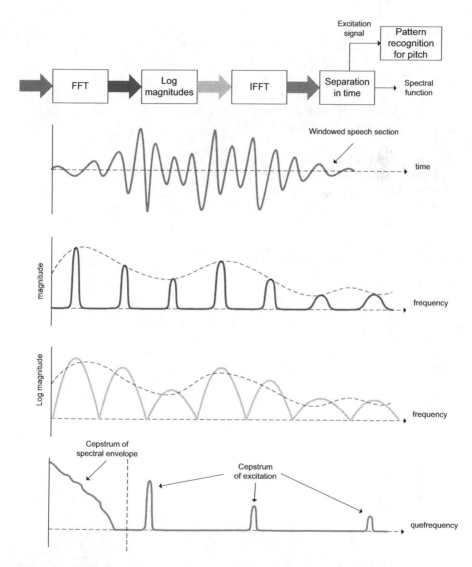

FIGURE 9.5: Cepstrum analysis. Reprinted with permission. From (Figure 20.1): Ben Gold, Nelson Morgan, and Dan Ellis. *Speech and Audio Signal Processing: Processing and Perception of Speech and Music.* John Wiley & Sons, 2011 [3].

Listing 9.1: Cepstrum Test Code

```
1  [y,fs] = audioread('JFK_1.wav');
2  x = y(:,2);
3
4  frameLength = 20;
5  nsample = round(frameLength * fs / 1000);
6  window = eval(sprintf('%s(nsample)','hamming'));
7  pos = 27343;
8  frame = x(pos:pos+nsample-1);
9  time=(0:length(frame)-1)/fs;
10 frameW = frame .* window;
11
12 Y = fft(frameW, nsample);
13 hz10000=10000*length(Y)/fs;
14 f=(0:hz10000)*fs/length(Y);
15 dbY10K = 20*log10(abs(Y(1:length(f)))+eps);
16 logY = log(abs(Y));
17 Cepstrum=ifft(logY);
18 Cepstrum10K = (Cepstrum(1:length(f)));
19
20 subplot(4,1,1);
21 plot(time,frame);
22 legend('Waveform'); xlabel('Time (s)'); ylabel('Amplitude');
23 subplot(4,1,2);
24 plot(time,frameW);
25 legend('Windowed Waveform'); xlabel('Time (s)'); ylabel('
       Amplitude');
26 subplot(4,1,3);
27 plot(f,dbY10K);
28 legend('Spectrum'); xlabel('Frequency (Hz)'); ylabel('Magnitude
       (dB)');
29 subplot(4,1,4);
30 plot(Cepstrum10K(2:end));
31 legend('Cepstrum'); xlabel('Quefrequency (s)'); ylabel('Level');
```

Listing 9.2: Cepstrum Truncation Test Code

```
1  [y,fs] = audioread('JFK_1.wav');
2  x = y(:,2);
3
4  frameLength = 20;
5  nsample = round(frameLength * fs / 1000);
6  window = eval(sprintf('%s(nsample)','hamming'));
7  pos = 27343;
8  frame = x(pos:pos+nsample-1);
```

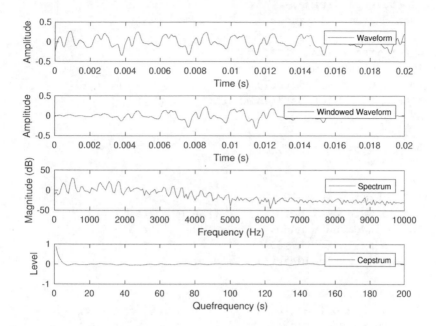

FIGURE 9.6: Cepstrum analysis: Generated from Listing 9.1.

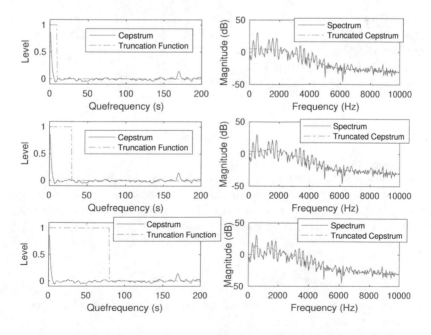

FIGURE 9.7: Cepstrum truncation: Generated from Listing 9.2.

```
 9  time=(0:length(frame)-1)/fs;
10  frameW = frame .* window;
11
12  Y = fft(frameW, nsample);
13  hz10000=10000*length(Y)/fs;
14  f=(0:hz10000)*fs/length(Y);
15  dbY10K = 20*log10(abs(Y(1:length(f)))+eps);
16  logY = log(abs(Y));
17  C=ifft(logY);
18  C10K = (C(1:length(f)));
19
20
21  thisC = C; trunNo = 10;
22  trunFn10 = zeros(length(f),1); trunFn10 (1:10) = 1;
23  thisC(trunNo:end-trunNo) = 0;
24  iC = exp(real(fft(thisC)));
25  dbiC10K10 = 20*log10(abs(iC(1:length(f)))+eps);
26
27  thisC = C; trunNo = 30;
28  trunFn30 = zeros(length(f),1); trunFn30 (1:30) = 1;
29  thisC(trunNo:end-trunNo) = 0;
30  iC = exp(real(fft(thisC)));
31  dbiC10K30 = 20*log10(abs(iC(1:length(f)))+eps);
32
33  thisC = C; trunNo = 80;
34  trunFn80 = zeros(length(f),1); trunFn80 (1:80) = 1;
35  thisC(trunNo:end-trunNo) = 0;
36  iC = exp(real(fft(thisC)));
37  dbiC10K80 = 20*log10(abs(iC(1:length(f)))+eps);
```

9.3.1 Properties of Cepstrum Analysis

The frequency transform of a speech signal (FFT) shows the harmonics of the source modulated by the resonances of the filter. The log FFT is, therefore, a sum of the source comb structure and the filter resonance peaks. A subsequent inverse frequency transform (IFFT) is able to separate the resonant bumps and the regular fine structure. In the FFT domain

- Filtering $\rightarrow$ Liftering

- Frequency $\rightarrow$ Quefrequency

Selecting the Cepstrum transform values below a threshold separates the filter and source.

FIGURE 9.8: Texas Instruments' "Speak and Spell" toy which utilises LPC for speech synthesis.

9.4 LPC Features

LPC is an acronym for Linear Predictive Coding. LPC has had a crucial role in speech analysis, synthesis and feature extraction for speech recognition. Figure 9.8 shows the Texas Instruments' "Speak and Spell" toy which utilises an LPC model for the synthesis of speech.

LPC aims to define an all-pole digital filter that fits a given signal. LPC features have been successfully applied to ASR as they are excellent at modelling the spectral envelope of the filter in the source-filter model. LPC analysis of speech features has the following characteristics:

- Models the resonations of the spectral envelope

- Extremely compact mathematical representation (all-pole filter)

- One pair of poles per resonation (formant)

- Good representation for most speech

    - However, nasals could be better represented with additional zeros (as there is usually a gap in the spectral envelope of nasals).

9.4.1   LPC Definitions

LPC analysis is defined as an autoregressive model. Autoregressive models are used in many areas of signal processing, e.g., image texture representation. The spectral response of the signal is modelled as the product of an all-pole filter and the spectral response of an error signal. This is a natural representation of the source filter paradigm. $E(z)$, the error signal can, therefore, be considered as the source excitation and the all-pole filter as the vocal tract filter. Similarly to spectrogram and Cepstrum analysis for speech signals, LPC analysis is implemented every 10 to 20ms. The justification for this is that it is assumed that the statistical properties of the filter aspect of a speech signal are approximately stationary over these timescales.

Defining:

- $s[n]$ as the signal (and $S(z)$ is its Z transform)

- $e[n]$ as the error signal (and $E(z)$ is its Z transform)

- $a_{1...p}$ are the filter parameters (for an all pole filter of length p)

The initial step for LPC analysis is to model the input signal as a linear combination of prior outputs and an error signal:

$$s[n] = e[n] + \sum_{j=1}^{P} a_j s[n-j].$$ (9.5)

Taking the Z-transform of both sides leads to:

$$S(z) = \underbrace{\frac{1}{\left(1 - \sum_{j=1}^{P} a_j z^{-j}\right)}}_{\text{All pole filter}} \underbrace{E(z)}_{\text{Excitation}}$$

(9.6)

9.4.2   LPC Spectral Model Filter

The goal of LPC spectral model filter analysis is to minimise $\sum_{n} e^2[n]$: the sum of the squared error. [for simplification, the signs of the weights are changed, i.e., $\alpha_k = -a_k$].

The expression for the error (given the above definition) is:

$$e[n] = s[n] + \sum_{k=1}^{P} \alpha_k s[n - k]. \tag{9.7}$$

By setting $\alpha_0 = 1$ this expression can be further simplified to:

$$e[n] = \sum_{k=0}^{P} \alpha_k s[n - k]. \tag{9.8}$$

The measure to minimise $\sum_{n} e^2[n]$ can now be expressed as follows:

$$\sum_{n} e^2[n] = \sum_{n} \left(\sum_{k=0}^{P} \alpha_k s[n-k] \right)^2, \tag{9.9}$$

$$= \sum_{n} \left(\sum_{i=0}^{P} \alpha_i s[n-i] \right) \left(\sum_{j=0}^{P} \alpha_j s[n-j] \right), \tag{9.10}$$

$$= \sum_{i=0}^{P} \sum_{j=0}^{P} \alpha_i \left[\sum_{n} s[n-i]s[n-j] \right] \alpha_j, \tag{9.11}$$

$$= \sum_{i=0}^{P} \sum_{j=0}^{P} \alpha_i c_{ij} \alpha_j, \tag{9.12}$$

where $c_{ij} = \sum_{n} s[n-i]s[n-j]$. It should be noted that $c_{ij} = c_{ji}$.

The goal is to minimise $\sum_{n} e^2[n]$ denoted as E. Using elementary calculus, the partial differentials of E with respect to each parameter (α_i) are set to zero:

$$\frac{\partial E}{\partial \alpha_k} = 0, \quad k = 1, 2, \ldots, P \tag{9.13}$$

$$\frac{\partial E}{\partial \alpha_k} = \underbrace{\sum_{j=0}^{P} c_{kj} \alpha_j}_{i = k} + \underbrace{\sum_{i=0}^{P} \alpha_i c_{ik}}_{j = k} = 2 \sum_{i=0}^{P} c_{ki} \alpha_i = 0, \quad k = 1, 2, \ldots, P$$

$$\tag{9.14}$$

$$2\sum_{i=0}^{P} c_{ki}\alpha_i = 0, \qquad k = 1, 2, \ldots, P \quad \text{Note: } c_{ij} = c_{ji} \tag{9.15}$$

Since $\alpha_0 = 1$

$$\sum_{i=1}^{P} c_{ki}\alpha_i = -c_{k0}, \qquad k = 1, 2, \ldots, P \tag{9.16}$$

as $a_k = -\alpha_k$

$$\sum_{i=1}^{P} c_{ki}a_i = c_{k0}, \quad k = 1, 2, \ldots, P \tag{9.17}$$

The next key step is to rearrange (9.17) into matrix form:

$$\begin{pmatrix} c_{11} & c_{12} & \cdots & c_{1P} \\ c_{21} & c_{22} & \cdots & c_{2P} \\ \vdots & \vdots & \ddots & \vdots \\ c_{P1} & c_{P2} & \cdots & c_{PP} \end{pmatrix} \begin{pmatrix} a_1 \\ a_2 \\ \vdots \\ a_P \end{pmatrix} = \begin{pmatrix} c_{01} \\ c_{02} \\ \vdots \\ c_{0P} \end{pmatrix} \tag{9.18}$$

Recalling that $c_{ij} = c_{ji}$ and $c_{ij} = \sum\limits_{n} s[n-i]s[n-j]$ the values of c_{ij} can be expressed reducing redundancy as:

$$c_{ij} = r_{|i-j|}. \tag{9.19}$$

This follows from the following insight:

$$c_{ij} = \sum_{n} s[n]s[n + (i-j)] = c_{0,(i-j)}, \tag{9.20}$$

i.e., each element c_{ij} depends just on the difference $|i - j|$. Therefore each value c_{ij} is replaced by $r_{|i-j|}$ calculated through 9.20. The key matrix relation (9.18) is now more compactly represented as

$$\mathbf{R}\vec{a} = \vec{R} \tag{9.21}$$

Where $\mathbf{R}$, $\vec{a}$ and $\vec{R}$ can be represented as follows

$$
\begin{pmatrix}
r_0 & r_1 & r_2 & \cdots & r_{p-1} \\
r_1 & r_0 & r_1 & \cdots & r_{p-2} \\
r_2 & r_1 & r_0 & \cdots & r_{p-3} \\
\vdots & \vdots & \vdots & \ddots & \vdots \\
r_{p-1} & r_{p-2} & r_{p-3} & \cdots & r_0
\end{pmatrix}
\begin{pmatrix}
a_1 \\ a_2 \\ a_3 \\ \vdots \\ a_p
\end{pmatrix}
=
\begin{pmatrix}
r_1 \\ r_2 \\ r_3 \\ \vdots \\ r_p
\end{pmatrix}
$$

$$
\mathbf{R} =
\begin{pmatrix}
r_0 & r_1 & r_2 & \cdots & r_{p-1} \\
r_1 & r_0 & r_1 & \cdots & r_{p-2} \\
r_2 & r_1 & r_0 & \cdots & r_{p-3} \\
\vdots & \vdots & \vdots & \ddots & \vdots \\
r_{p-1} & r_{p-2} & r_{p-3} & \cdots & r_0
\end{pmatrix}
\quad
\vec{a} =
\begin{pmatrix}
a_1 \\ a_2 \\ a_3 \\ \vdots \\ a_p
\end{pmatrix}
\quad
\vec{R} =
\begin{pmatrix}
r_1 \\ r_2 \\ r_3 \\ \vdots \\ r_p
\end{pmatrix}
\tag{9.22}
$$

Due to the definition of elements c_{ij} and r, the highlighted diagonals shown in (9.22) contain equal values. Therefore $\mathbf{R}$ is a "Toeplitz" Matrix (equals its own transpose). The final step in order to obtain the values of $a_{1\ldots p}$ is to invert (9.21) thus:

$$
\vec{a} = \mathbf{R}^{-1}\vec{R}. \tag{9.23}
$$

This method of obtaining $\vec{a}$ is known as the "Autocorrelation Method". A summary of this method is given in Box 9.4.1.

Box 9.4.1: The Autocorrelation Method of LPC Analysis

From a signal $s[n]$ the weights $a_{1\ldots p}$ of an LPC filter (from (9.5)),

$$
\frac{1}{\left(1 - \sum_{j=1}^{P} a_j z^{-j}\right)}
$$

Can be obtained as follows

- Decide the order P of the filter.

- Obtain the "autocorrelation values" $r_{0\ldots P}$ from (9.19) and (9.20) [a]

- Form $\mathbf{R}$ and $\vec{R}$ defined in (9.22) from the $r_{0\ldots P}$ values.

- Use $\vec{a} = \mathbf{R}^{-1}\vec{R}$ to obtain $\vec{a}$ and therefore the weights of the LPC filter defined above.

---

[a] Autocorrelation values are often most effectively calculated using an FFT.

9.4.3 LPC: Autocorrelation Method: Example

Find a_1 and a_2 of a second order LPC filter defined by equation:

$$S(z) = \frac{1}{\left(1 - \sum_{j=1}^{2} a_j z^{-j}\right)} E(z).$$

Given the autocorrelation coefficients $r_0 = 1.0$, $r_1 = 0.5$ and $r_2 = 0.2$:

$$\mathbf{R} = \begin{pmatrix} r_0 & r_1 \\ r_1 & r_0 \end{pmatrix} = \begin{pmatrix} 1 & 0.5 \\ 0.5 & 1 \end{pmatrix} \qquad \vec{R} = \begin{pmatrix} r_1 \\ r_2 \end{pmatrix} = \begin{pmatrix} 0.5 \\ 0.2 \end{pmatrix} \qquad \boxed{\vec{a} = \mathbf{R}^{-1}\vec{R}}$$

$$\mathbf{R}^{-1} = \begin{pmatrix} 1 & 0.5 \\ 0.5 & 1 \end{pmatrix}^{-1} \qquad \vec{a} = \begin{pmatrix} 1.333 & -0.667 \\ -0.667 & 1.333 \end{pmatrix} \begin{pmatrix} 0.5 \\ 0.2 \end{pmatrix} = \begin{pmatrix} 0.533 \\ -0.067 \end{pmatrix}$$

Therefore $a_1 = 0.533$ and $a_2 = -0.067$.

9.4.4 LPC Analysis for Speech–Justification

From the theoretical analysis of a tube with one end open; complex poles will occur on the unit circle at frequencies given by:

- $F_n = (2n + 1)c/4l, n = 0, 1, 2, \ldots$

- L = 17cm, c =344m/s

- Therefore: one resonance per kHz

- One pole pair needed per resonance

- One pole pair needed for driving waveform

- Therefore 2*(BW+1) LPC coefficients required where BW is the speech bandwidth in kHz

9.4.5 Spectrum Estimation of Windowed Vowel Signal Using LPC

Figure 9.9 illustrates the LPC parameter extraction and modelling of a windowed vowel signal.

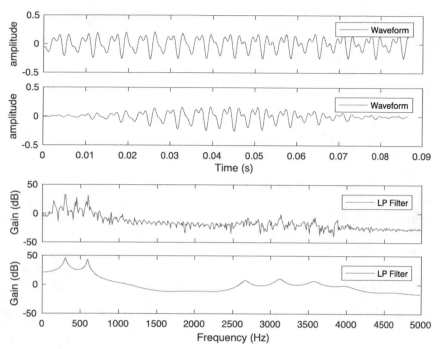

FIGURE 9.9: LPC analysis of a speech spectral envelope (generated by code in Listing 9.3).

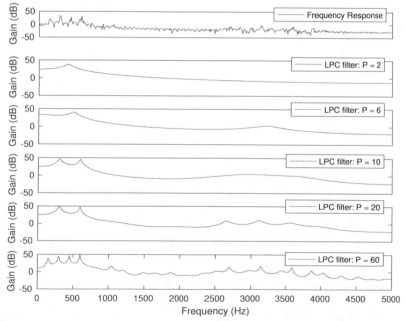

FIGURE 9.10: LPC analysis of a speech spectral envelope using various values of P.

Listing 9.3: LPC extraction code for a vowel based speech file

```
1  [x,fs]=audioread('paulhill2.wav',[24020 25930]);
2  x=resample(x,10000,fs); fs=10000;
3  t=(0:length(x)-1)/fs;
4  subplot(4,1,1);
5  plot(t,x);
6  legend('Waveform');xlabel('Time (s)');ylabel('amplitude');
7  x1 = x.*hamming(length(x));
8  subplot(4,1,2);
9  plot(t,x1);
10 legend('Waveform');xlabel('Time (s)');ylabel('amplitude');
11 ncoeff=20; % rule of thumb for formant estimation
12 preemph = [1 0.63];
13 x1 = filter(1,preemph,x1);
14 a=lpc(x1,ncoeff);
15 [h,f]=freqz(1,a,512,fs);
16 subplot(4,1,4);
17 plot(f,20*log10(abs(h)+eps));
18 legend('LPC Filter'); xlabel('Frequency (Hz)');ylabel('Gain (dB)
       ');
19
20 subplot(4,1,3);
21 outfft = fft(x, 1024);
22 plot(f,20*log10(abs(outfft(1:512))+eps));
23 legend('Frequency Response'); xlabel('Frequency (Hz)');ylabel('
       Gain (dB)');
```

9.4.6   LPC - Speech Spectra for Different Model Orders

Figure 9.10 shows how the variation of LPC order (using the parameter P) varies how well the LPC spectral envelope matches the spectrum/spectral envelope of the input signal.

- If the model order is too small

  - Difficult to effectively represent formants

- If the model order is too large

  - Unnecessary modelling of spectral harmonics
  - Low error, ill conditioned matrices

9.5 Feature Extraction for ASR

Commonly, feature representations are extracted every 10ms over a window of 20 or 30ms for speech analysis (similarly to LPC analysis). The aim is to find features that are stable for different examples of the same sounds in speech, despite differences in the speaker or the speaker's environment. Cepstrum and LPC analysis can be used for ASR. However, they do not integrate the full range of perceptually invariant components that are needed to generate the most effective features for speech recognition. Common perceptually invariant based features are:

- **MFCC** Mel Frequency Cepstral Coefficients

- **PLP** Perceptual Linear Prediction [1]

These currently popular choices were developed by the slow historical evolution of feature extraction techniques and are based as much on successful results as principled definition. These two types of perceptually invariant ASR features share many elements and are based on Cepstral approaches integrating perceptual features.

9.5.1 Short-Term Frequency Analysis

As is common with many frequency-based analysis audio systems ASR feature vectors are extracted using the common structure of the extraction of power estimates from FFT coefficient magnitudes. This is achieved similarly to the structure of spectrogram analysis illustrated in Figure 9.11. This figure illustrates that the formants change with different vowels. However, the position of these vowels (in terms of frequency) will vary according to the sex, age and physical characteristics of the speaker.

9.5.2 Cepstrum for Speech Recognition?

Recalling that the Cepstrum transform is defined as: IFFT $(\log |\text{FFT}(s[n])|)$. The Cepstrum is better than directly using short-term spectrograms as it is firstly decorrelated and secondly has a more compact representation. This is due to the truncation of the Cepstrum to form a representation of the filter characteristics of the human voice. However it is not based on perceptual principles, therefore PLP [4] and MFCC [2] methods use the Cepstrum as their basis, modified for invariant representations.

---

[1]A more recent variant of PLP is RASTA-PLP utilised by the HTK toolkit.

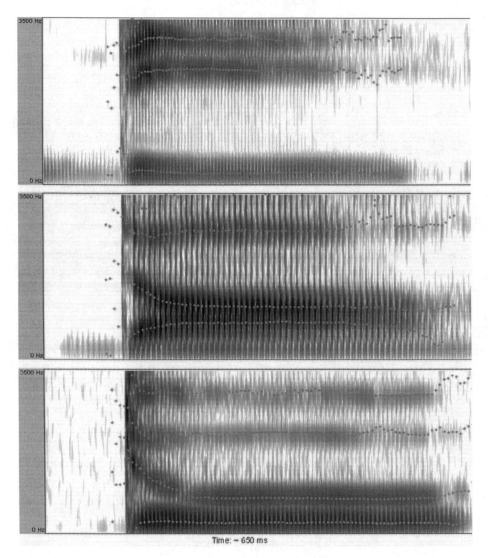

FIGURE 9.11: Spectrograms of the syllables "dee," "dah," and "doo." This image shows the first four formants highlighted by dots [1]. Copyright © Jonas Kluk, licensed under CC BY 3.0 https://creativecommons.org/licenses/by/3.0/.

9.6 Perceptually-Based Features: PLP and MFCC Features

PLP and MFCC features are usually extracted over "frames" of the order of 20ms. PLP and MFCC feature extraction methods share many components. A comparison of the feature extraction methods of the Cepstrum, PLP and MFCC features is shown in Figure 9.12. PLP and MFCC methods share the first five stages of feature extraction:

- **Stage 1**: The power spectrum for a windowed "frame" signal is calculated

- **Stage 2**: The power spectra are integrated using overlapping critical band filter responses (Figure 9.15 and 9.14)

- **Stage 3**: Pre-emphasis equalisation to create equal loudness across all frequencies

- **Stage 4**: Compress the spectral amplitudes
    - PLP takes cube root
    - MFCC takes log

- **Stage 5**: Inverse Discrete Fourier Transform
    - Only real inverse needed

- **Stage 6**: Spectral smoothing
    - PLP uses an autoregressive (LPC type) estimation of the compressed critical band spectrum
    - MFCC Cepstral truncation

- **Stage 7**: PLP only, the autoregressive coefficients are converted to Cepstral variables

9.6.1 Pre-Emphasis/Equal Loudness Filters

Most feature extraction methods for ASR (e.g., MFCC and PLP) incorporate a model of human frequency perception in order to perceptually normalise the features in terms of frequency content.

MFCC Pre-Emphasis Filter

The pre-emphasis filter used by most MFCC implementations (e.g., HTK [6]) use a simple single coefficient auto-regressive filter as defined in Equation (9.24). Although a is left as a tunable parameter it is usually set to be approximately 0.97 in HTK. This gives a very basic high pass filter:

$$y[n] = x[n] - ax[n-1]. \tag{9.24}$$

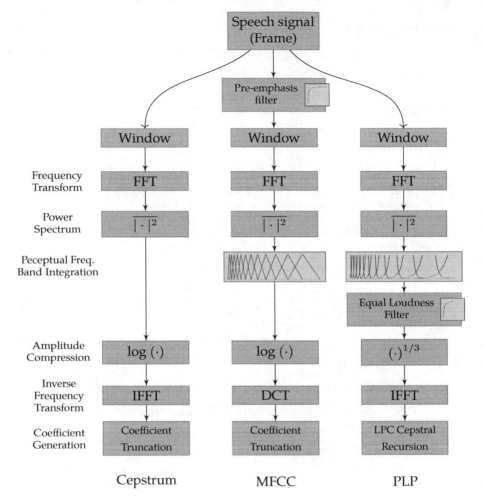

FIGURE 9.12: Cepstrum, PLP and MFCC feature extraction methods.

PLP Equal Loudness Curves

A pre-emphasis filter has been designed for PLP implementations. The most common form was proposed by Hermansky [4]:

$$E\left(\omega\right) = \frac{\left(\omega^2 + 56.8 \cdot 10^6\right) \cdot \omega^4}{\left(\left(\left(\omega^2 + 6.3 \cdot (10^6)\right)^2\right) \cdot \left(\omega^2 + 0.38 \cdot (10^9)\right)\right)}, \qquad (9.25)$$

where ω is defined as the angular frequency ($\omega = 2\pi \cdot f$). This equation was further extended within the same paper [4] for higher frequencies:

$$E\left(\omega\right) = \frac{\left(\omega^2 + 56.8 \cdot 10^6\right) \cdot \omega^4}{\left(\omega^2 + 6.3 \times 10^6\right)^2 \cdot \left(\omega^2 + 0.38 \times 10^9\right) \cdot \left(\omega^6 + 9.58 \times 10^{26}\right)}. \qquad (9.26)$$

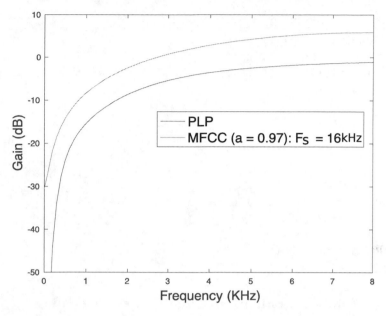

FIGURE 9.13: Pre-emphasis filters of PLP [4] and MFCC [2] methods. The MFCC filter makes the assumption that F_s is 16kHz. This is the default for the HTK toolkit [6].

However, (9.25) is the most common form and has been adopted within the implementation of PLP features within HTK for example [6].

Figure 9.13 was created from MATLAB code shown in Listing 9.4.

Listing 9.4: MFCC Pre-emp Filter

```
1   f1 = 0:100:8000; w = 2*pi*f1;
2   %PLP Pre=Emphasis Filter Definition
3   Ew1 = ((w.^2+56.8*(10^6)).*(w.^4))./(((w.^2+6.3*(10^6)).^2).*(w
       .^2+0.38*(10^9)));
4
5   Ew1db = 20*log10(Ew1);
6   plot(f1./1000, Ew1db,'b');hold on;
7
8   % MFCC Pre-Emphasis Filter Definition
9   a = 1; b = [1 -0.97];
10  MELFS = 16000; %typical sampling rate for HTK MFCC
11  radiansW1 = 2*pi*f1/MELFS;
12  h = freqz(b,a,radiansW1);
13  plot(f1/1000,20*log10(abs(h)),'r');
```

9.6.2   Critical Band Integration: Perceptual Frequency Pooling

The linear indexing scale of the power spectral coefficients output from stage 2 is not aligned with auditory perception. In order to provide a more compact representation and more accurately align with frequency perception both the PLP and MFCC feature extraction methods utilise critical band integration. These critical bands are illustrated in Figures 9.14 (for PLP) and 9.15 (for MFCC). For each of these bands, the power spectral coefficients of the FFT are summed under each filter shape (triangular for MFCC filters in Figure 9.15 and more block-like filter shapes for PLP in Figure 9.14). This ensures that the integrated values are not only better aligned with frequency perception, but give a less redundant (and compact) representation compared to the original FFT spectral power coefficients.

Specifically, Figure 9.15 illustrates the MFCC filter shapes created by HTK [6] when specifying the number of filters as being 20. Within HTK, the MFCC bands were based on the Mel scale given by Equation (5.10) and the PLP bands were based on the Bark scale. Both scales are defined in Chapter 5.

9.6.3   Perceptual Loudness Compensation

After the critical band integration stage, both the PLP and MFCC methods perform a type of perceptual loudness compensation. The PLP method takes the cube root of the output of the critical band integration. This is an approximation of the power law of hearing as described in Chapter 5.

Conversely, a logarithm is taken of the integrated FFT magnitudes within the MFCC method. The justification of this was:

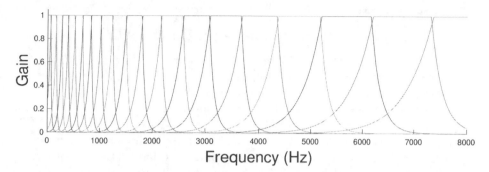

FIGURE 9.14: Critical band filters (based on the Bark scale) used to integrate the power spectral coefficients within the PLP: 20 bands, 16kHz maximum frequency (default choices for HTK analysis).

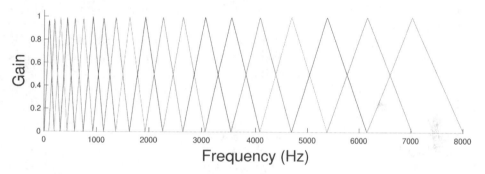

FIGURE 9.15: Critical band filters (based on the Mel scale) used to integrate the power spectral coefficients within the MFCC: 20 bands, 16kHz maximum frequency (default choices for HTK analysis)

- The logic of reducing a multiplication in the frequency domain to an addition (making separation of source and filter easier)

- Logarithm compresses the signal dynamic range

- Perceptual of level differences is logarithmic

- Frequency estimates become less sensitive to slight variations in input power

- Phase information is often not considered to be important for speech recognition

9.7   Practical Implementations of PLP and MFCC

The key parameter for both PLP and MFCC analysis is the number of critical bands to integrate with, the maximum analysis frequency, the final number of coefficients and the form of any derived coefficients.

There are numerous possible choices. However as a guide, the default number of critical bands within the MFCC implementation within the HTK toolkit is 20. Furthermore, within the HTK toolkit, the final number of coefficients is set to a default of 12 and the sampling frequency is set to 16kHz.

It should be noted that a well-used extension to PLP processing the so-called PLP-RASTA method, has become significantly popular recently (it is also implemented within the HTK toolkit).

The HTK coefficients typically used by the HTK toolkit use what are known as MFCC "deltas"(Δ) and MFCC "delta-deltas"($\Delta\Delta$), i.e., the change in MFCC coefficients and the change of the change in MFCC coefficients from frame to frame.

The following shows a typical extraction of MFCC features using the HTK toolkit: the MFCC features are the native features and the Del and Acc features are the MFCC "deltas"(Δ) and MFCC "delta-deltas"($\Delta\Delta$), respectively.

```
Source: example_file
Sample Bytes: 26 Sample Kind: MFCC_0
Num Comps: 13 Sample Period: 10000.0 us
Num Samples: 336 File Format: HTK
------------------- Observation Structure --------------------
x: MFCC-1 MFCC-2 MFCC-3 MFCC-4 MFCC-5 MFCC-6 MFCC-7
MFCC-8 MFCC-9 MFCC-10 MFCC-11 MFCC-12 C0 Del-1
Del-2 Del-3 Del-4 Del-5 Del-6 Del-7 Del-8
Del-9 Del-10 Del-11 Del-12 DelC0 Acc-1 Acc-2
Acc-3 Acc-4 Acc-5 Acc-6 Acc-7 Acc-8 Acc-9
Acc-10 Acc-11 Acc-12 AccC0
----------------------- Samples: 0->1 -------------------------
0: -14.514 -3.318 -5.263 -6.245 6.192 3.997 1.830
5.293 3.428 3.731 5.829 5.606 30.734 -0.107
-2.360 1.331 1.134 -0.423 -0.576 0.083 0.502
-1.587 -0.592 -0.372 -0.330 -0.171 0.256 0.120
-1.441 -0.226 -3.516 -0.144 -0.055 0.243 -0.039
1.197 0.275 -0.295 0.251
-------------------------- END ----------------------------
```

9.8   Generic Audio Features

9.8.1   Mid- and Short-Term Feature Audio Analysis

A very large number of features and feature types can be extracted from audio signals. Audio features can be specifically developed for speech analysis and recognition (as described in Section 9.6). However, within this section, generic audio features are described that can be used for more general audio signals (such as music, etc.) for characterisation and retrieval. Such generic audio features can be extracted on a short-term to mid-term basis (defined below).

9.8.2   Short Term Feature Extraction

Short term feature extraction is based on separating the input audio signal into overlapping or non-overlapping "frames". Typical frame/window lengths are approximately 10-30ms. This is because there will be an applicable number of pitch periods (for music, speech or vibration based audio) within that time length.

9.8.3   Medium Term Feature Extraction

Medium-term feature extraction is also based on separating the input audio signal into overlapping or non-overlapping "frames" that are in general much longer in time than short-term feature extraction (usually with frame lengths of between 1 second to 10 seconds).

9.8.4   Time Domain Energy Audio Features Example: Zero Crossing Rate

Many different frequency and time domain audio features have been used in the analysis of audio. The zero crossing rate is an example of such a feature. For a frame of analysis, the zero crossing rate is the number of times the signal crosses zero (i.e., changed either from positive to negative or from negative to positive):

$$\text{ZCR}(x) = \frac{1}{2(N-1)} \sum_{i=1}^{N-1} |\text{sign}(x[i]) - \text{sign}(x[i-1])|, \qquad (9.27)$$

where N is the number of samples of an audio file and the signum function is defined by:

$$\text{sign}(x[i]) = \begin{cases} 1, & \text{if } x[i] > 0 \\ 0, & \text{if } x[i] = 0 \,. \\ -1, & \text{if } x[i] < 0 \end{cases} \quad (9.28)$$

The number of zero crossings is an excellent and simple feature that is easy to understand and extract. However, the zero crossing rate is often very sensitive to irrelevant changes in an audio signal leading to the zero crossing rate having limited usefulness in these circumstances. These irrelevant changes are often typified by low amplitude "noise", fluctuations in DC offset or existing audio "jitter" that changes the zero crossing rate significantly and is independent of the key parts of an audio signal.

9.9   ASR Features: Summary

- Feature extraction is required to provide a compact and invariant representation for ASR

    - Physiological based filters for audio analysis

    - Compact representation required: Features

    - Source-filter theory leads to analysis using FFTs, Linear prediction and Cepstral analysis

- Features for general audio analysis can be achieved over short/medium and long time frames using features such as the zero crossing rate.

9.10   Exercises

Exercise 9.1
Create some MATLAB code to generate the zero crossing for a short time sliding analysis window and plot the output on a figure with the actual waveform.

Exercise 9.2
Create some MATLAB code to generate and plot a series of pre-emphasis filters where the a parameter of Equation (9.24) is from the following list {0.7, 0.8, 0.9}.

Bibliography

[1] https://upload.wikimedia.org/wikipedia/commons/thumb/7/75/Spectrograms_of_syllables_dee_dah_doo.png/.

[2] S. Davis and P. Mermelstein. Comparison of parametric representations for monosyllabic word recognition in continuously spoken sentences. *IEEE Transactions on Acoustics, Speech, and Signal Processing*, 28(4):357–366, 1980.

[3] B. Gold, N. Morgan, and D. Ellis. *Speech and Audio Signal Processing: Processing and Perception of Speech and Music*. John Wiley & Sons, 2011.

[4] H. Hermansky. Perceptual linear predictive (plp) analysis of speech. *The Journal of the Acoustical Society of America*, 87(4):1738–1752, 1990.

[5] A.V. Oppenheim. *Discrete-Time Signal Processing*. Pearson Education India, 1999.

[6] S.J. Young, D. Kershaw, J. Odell, D. Ollason, V. Valtchev, and P. Woodland. *The HTK Book Version 3.4*. Cambridge University Press, 2006.

10

HMMs, GMMs and Deep Neural Networks for ASR

CONTENTS

> Everything we see hides another thing, we always want to see what is hidden by what we see.
>
> René Magritte

This chapter firstly introduces Hidden Markov Models (HMMs) as a key underlying method for the analysis of speech. HMMs have been extremely important for current and historical ASR systems. Until fairly recently the state of the art performance of ASR systems was achieved through the combination of HMM and Gaussian Mixture Models (GMMs). More recent models have replaced GMMs with Deep Neural Networks (DNNs). For large training datasets, the current state of the art ASR systems have moved away from using HMMs and now directly use DNNs such as RNNs in so-called end-to-end DNNs for ASR.

10.1   Hidden Markov Models

Markov and Hidden Markov Models (HMMs) are methods to find and define patterns over space and time. HMMs are applicable to speech recognition as they are able to take into account the timing variability between speakers (and also the same speaker). Other motivations for using HMMs include:

- Many real-world processes are sequential (the process unfolds over time) and the event producing the output is hidden.

- Internal states of a system (such as the internal states of a voice producing speech) are often hidden.

10.1.1   ASR Using Hidden Markov Models (HMMs)

The mathematical principles of HMMs were introduced by Baum et al. in the 1960s and 1970s [4, 13].[1] Following on from their mathematical development they were first applied to ASR within a team at IBM from 1970 onwards (Baker, Jelinek, Bahl, Mercer,. . .) [3, 9]. They have been used in state of the art ASR systems from the mid 1980s until the present. Although presently, Deep Networks have been used for state of the art speech recognition systems they are often combined with HMMs for effective ASR. HMMs are also used for:

- Text processing
  - Parsing raw records into structured records
- Bioinformatics
  - Protein sequence prediction
- Financial Applications
  - Stock market forecasts (price pattern prediction)
  - Comparison shopping services
- Image Processing

---

[1]Although there was an initial formulation described even earlier by Ruslan L. Stratonovich.

10.2 Mathematical Framework of HMMs

In order to understand the mathematical framework of HMMs, firstly, simple deterministic patterns and next Markov processes are examined. From these building blocks an HMM will be defined and illustrated.

10.2.1 Deterministic Patterns

Deterministic patterns are simple sequential patterns whereby one state changes to the next state in a deterministic order, e.g.,traffic lights (see Figure 10.1).

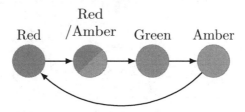

FIGURE 10.1: Deterministic patterns example: traffic lights (in the U.K.).

10.3 Non-Deterministic Patterns

Markov processes are non-deterministic, i.e., they do not always follow the same sequence of states. Each state transition is determined by the previous n-states giving an nth-order Markov process. We are mostly interested in first-order Markov processes. Figure 10.2 shows an example of a first order Markov process. In this example the "weather" can be in one of three observable states: {Sunny, Rainy, Cloudy}. State transitions are associated with a set of state transition probabilities. It is assumed that state transition probabilities are independent over time (this is often a gross assumption).

10.3.1 Markov Processes

A Markov Process is defined by:

- A set of states (e.g., sunny, cloudy and rainy)

- A state transition probability matrix $\mathbf{A}$

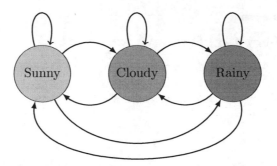

FIGURE 10.2: Non-deterministic pattern example: A first order Markov Process, the weather with three states on a given day: cloudy, rainy and sunny.

- An initial state probability vector **Π**

- An example of a Markov process is illustrated in Figure 10.2.

The probability of a sequence of states equals the product of the probability of the initial state and the subsequent state transitions. For the first order Markovian process illustrated in Figure 10.2 an example transition matrix **A** could be (Figure 10.3):

		Weather Tomorrow		
		sunny	cloudy	rainy
	sunny	0.5	0.25	0.25
Weather Today	cloudy	0.2	0.6	0.2
	rainy	0.3	0.15	0.55

FIGURE 10.3: Example state transition matrix **A** of the Markov process illustrated in Figure 10.2.

10.4   Hidden Markov Models (HMMs)

Markov models are not always appropriate for systems where the state of the system is not observable, i.e., when measuring something directly in order to determine a hidden state, e.g.

- using a barometer to determine the weather outside on a specific day

- using acoustic features to determine the hidden vocal state at any time

These are the type of problems that Hidden Markov Models are designed to solve. For HMMs, each state is not actually observed (i.e., it is hidden) and only a set of observations of the system are available (output).

10.4.1   Hidden Markov Models: Definition

An HMM is comprised of two sets of states:

- hidden states: the "true" states of a system. These "true" states may be described by a Markov process (e.g., the weather).

- observable states: the visible states (e.g., barometer readings).

A particular HMM is defined by a triple of parameters $\{\Pi, A, B\}$.

- Π the vector of the initial state probabilities.

- A the state transition matrix. Each probability depends only on the previous state (i.e., assume first-order Markov process).

- B the emission probability matrix. This is a matrix of the probability of each state occurring at the same time as an observation.

- A and B are assumed to be time independent. This is unrealistic in practice.

In order to extend a three state first order Markovian model as illustrated in Figure 10.2 to an HMM it is first assumed that original states are now hidden and there is a set of observed states associated with the hidden state at each time step. Such a model is illustrated in Figure 10.4.

10.5   Hidden Markov Models: Mathematical Definition

An HMM is mathematically defined as follows:

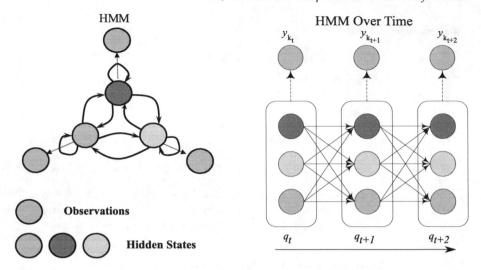

FIGURE 10.4: Hidden Markov Model (first order): q_t is the state at time t, y_{k_t} is the observed state at time t (where k_t is the index of the observed state at time t).

- $\lambda = \{\mathbf{\Pi}, \mathbf{A}, \mathbf{B}\}$ triple of parameters defining the HMM

- $\mathbf{\Pi} = \{\pi_i\}$ vector of the initial state probabilities

  $$\pi_i = P[q_1 = s_i] \quad 1 \leq i \leq N$$

- $\mathbf{A} = \{a_{ij}\}$ the state transition matrix

  $$a_{ij} = P[q_{t+1} = s_j | q_t = s_i] \quad 1 \leq i, j \leq N$$

- $\mathbf{B} = \{b_{ij}\}$ the emission probability matrix

  $$b_{ij} = P[y_i | q_t = s_j] \quad 1 \leq i \leq M,\ 1 \leq j \leq N$$

- M is the number of observation states

  - The individual observation states are denoted as $Y = \{y_1, y_2, \ldots, y_M\}$
  - The observed state at time t is y_{k_t}. Where $k_t \in \{1, \ldots M\}$ is the index of the observed state at time t

- N is the number of hidden states

  - The individual hidden states are denoted as $S = \{s_1, s_2, \ldots, s_N\}$
  - The hidden state at time t is $q_t \in S$

10.5.1 Hidden Markov Models: Example

Figures 10.4 and 10.5 illustrate a three hidden state HMM example. This example being a simple extension of the three state first order Markovian process illustrated in Figure 10.2. The states are still: rainy, sunny, cloudy. However, it is now assumed that they are not observable. Associated with the hidden state at each time step there is one of a set of observation states. In this case, there are three observation states. These states are the measurement of barometric pressure from a barometer. This measurement is quantised into three discrete states: high, medium and low (the pressure level measured by the barometer). If the weather states are not observable it can be assumed that the observer of the system is locked in a cupboard with a barometer (or some other bizarre alternative).

Π vector of initial State probabilities		A state transition matrix				B the emission probability matrix			
			Weather Tomorrow				**Observation**		
			sunny	cloudy	rainy		High	Medium	Low
sunny	0.5	sunny	0.5	0.25	0.25	sunny	0.7	0.2	0.1
cloudy	0.2	**Weather Today** cloudy	0.2	0.6	0.2	**Weather Today** cloudy	0.3	0.6	0.1
rainy	0.3	rainy	0.3	0.15	0.55	rainy	0.3	0.5	0.2

FIGURE 10.5: Example HMM with three hidden states and three observable states.

10.6 HMM Problems

There are three problems associated with HMMs:

Evaluation: Calculating the probability of an observed sequence of observations given an HMM model. When we have multiple HMM models we need to evaluate which model is the most likely to have generated (in terms of probability) a set of observations, i.e., in the above example, we could train the model to be "summer" or "winter" and then find the probability associated with each model (the probability that the observations have been generated by the model).

Decoding: Finding the most probable sequence of hidden states given a set of observations.

Learning: Generating the parameters of an HMM from a sequence of training observations.

10.7 Hidden Markov Models: Evaluation

HMM evaluation is the process of finding how probable an HMM model generated a set of observations, i.e., we want to find the probability of an observed sequence given an HMM defined by the triple $\lambda = \{\mathbf{\Pi}, \mathbf{A}, \mathbf{B}\}$. Given the weather and barometer example (Figure 10.5), a set of observations is illustrated for an example shown in Figure 10.6, i.e., the set of barometer observations is (high, high, low) on consecutive days. On each of these days, the weather may be rainy, cloudy or sunny (but we cannot observe these states). The potential (although not observed) change of hidden state is shown as a trellis also illustrated within Figure 10.6.

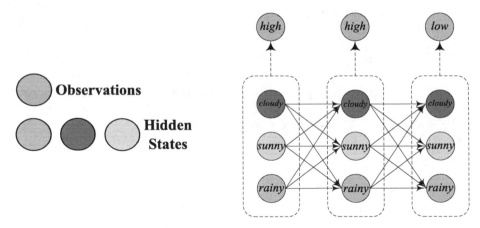

FIGURE 10.6: A first order HMM with three hidden states and three observable states.

The probability of the observation sequence can be calculated by summing an exhaustive set of probabilities of all the possible hidden states given the observed sequence. For the example illustrated in Figure 10.6, there would be $3^3 = 27$ possible different weather sequences, and so the probability that this sequence of observations (high,high,low) is given by:

$$P(\text{high,high,low}|\text{HMM}) = \quad P(\text{high,high,low}|\text{sunny,sunny,sunny}) +$$
$$P(\text{high,high,low}|\text{sunny,sunny,cloudy}) +$$
$$P(\text{high,high,low}|\text{sunny,sunny,rainy}) +$$
$$\cdots +$$
$$\cdots +$$
$$P(\text{high,high,low}|\text{rainy,rainy,rainy}). \quad (10.1)$$

This exhaustive method is obviously computationally expensive.

10.7.1 HMM Evaluation: Forward Algorithm

The computational expense of calculating Equation (10.1) can be reduced using recursion. This method is known as the Forward Algorithm. Partial probabilities denoted $\alpha_t(j)$ are associated with all the hidden states at all time steps (of the observation sequence). Partial probabilities $\alpha_t(j)$ represent the probability of getting to a particular state, j, at time t. Partial probabilities α are calculated using the following method:

- To calculate the probability of getting to a state through all paths, we can calculate the probability of each path to that state and sum them.

$$\alpha_t(j) = \quad P(\text{observation at time } t| \text{ hidden state is } j \text{ at time } t) \times$$
$$P(\text{all paths to state } j \text{ at time } t) \quad (10.2)$$

- The number of paths needed to calculate $\alpha_t(j)$ increases exponentially as the length of the observation sequence increases but the α's at time $t-1$ give the probability of reaching that state through all previous paths

  - We can, therefore, calculate the α's at time t in terms of those at time $t-1$

- Given an input set of observation states of length T

  - $Y^{(k)} = y_{k_1}, y_{k_2}, \ldots, y_{k_T}$ e.g. high, high, low

- Calculating α for all states at $t > 1$

  - The probability of getting to a particular state at time t, $\alpha_t(j)$ can be calculated from summing the probability of getting to this state from all the previous states.

The expression for calculating each partial probability $\alpha_{t+1}(j)$ from the partial probabilities at time t is:

$$\alpha_{t+1}(j) = b_{k_{t+1}j} \sum_{i=1}^{N} \alpha_t(i)a_{ij}, \tag{10.3}$$

i.e., it is the sum of the products of the relevant state transition matrix a_{ij} elements with the previous partial probabilities $\alpha_t(i)$ multiplied by the relevant element of the observation matrix $b_{k_{t+1}j}$. This iterative step is illustrated in Figure 10.7

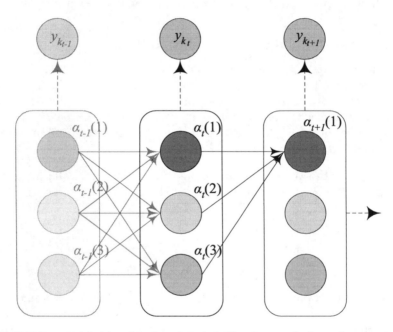

FIGURE 10.7: Hidden Markov Model: The forward algorithm example.

Forward Algorithm: Calculating the Overall Probability: Equation (10.3) shows how all the partial probabilities can be calculated when the previous timestep partial probabilities exists. In order to initialise the process the first partial probabilities at time $t = 1$ are calculated using the initial probability vector as follows:

$$\alpha_1(j) = b_{k_1 j} \cdot \pi(j). \tag{10.4}$$

And finally the probability of the observed sequence given an HMM model $P(Y^{(k)}|\lambda)$ is achieved by summing the final partial probabilities at time $t = T$ (i.e., the last time step), i.e.,

$$P(Y^{(k)}|\lambda) = \sum_{j=1}^{N} \alpha_T(j) \tag{10.5}$$

HMM Forward Algorithm: Example

Figures 10.7 and 10.8 show an example of the forward algorithm used with a three hidden state HMM. This HMM has three hidden states corresponding to three phones and three observation states (first formant frequencies F_1, quantised into three possible states: {low,medium,high}). Three time steps are shown in the bottom of the figure and are associated with a given observation sequence: high,high,low. The partial probabilities have been calculated for the first two time steps and the calculations leading to the values located by the question marks are shown to the right of the question marks. These calculations are based on Equation (10.3). The final probability of the HMM is calculated from (10.5) and is shown on the bottom line. This example shows that the calculation of each partial probability is a vector product of the previous partial probabilities, the relevant column in the state transition matrix (**A**) multiplied by the relevant element of the observation matrix (**B**).

10.8 Hidden Markov Models: Decoding

HMM decoding is the process of finding the most probable sequence of hidden states within the model. As with HMM evaluation the most simple method conceptually is the exhaustive solution:

- Find all the possible sequences of hidden states. Calculate the probability of each of the possible sequences for the sequence of observations.

- Find the most probable sequence of hidden states by finding the maximum of P(observed sequence | hidden state combination).

As with exhaustive HMM evaluation, this method is very computationally expensive. Reduced complexity is achieved using recursion: the Viterbi Algorithm.

10.8.1 HMM Decoding: The Viterbi Algorithm

The computational expense of the exhaustive method of HMM decoding described above can be reduced using recursion. This recursive method is known as the Viterbi Algorithm. Partial probabilities denoted $\delta_t(j)$ are associated with all the hidden states at all time steps (of the observation sequence).

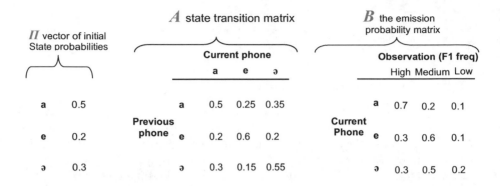

Partial Probabilities α

	high	high	low	
a	0.35	0.1498	?	(0.1498×0.5+0.0411×0.2+0.0552×0.3)×0.1 = 0.009968
e	0.06	0.0411	?	(0.1498×0.25+0.0411×0.6+0.05520.15)×0.1 = 0.007039
ə	0.09	0.0552	?	(0.1498×0.35+0.0411×0.2+0.0552×0.55)×0.2 = 0.018202

Total probability = 0.018202 + 0.007039 + 0.009968 = 0.035209

FIGURE 10.8: Example forward algorithm implementation. Equation (10.3) for right-hand column and Equation (10.5) for total probability calculation.

Partial probabilities $\delta_t(j)$ represent the probability of getting to a particular state, j, at time t.

- Partial probabilities for HMM decoding

  - **A partial probability** $\delta_t(i)$ is the maximum probability of all the sequences ending at state i at time t. The **partial best path** is the sequence of hidden states which leads to this maximum.
  - Unlike the partial probabilities in the forward algorithm, δ is the probability of the one (most probable) path to the state.

- Viterbi Algorithm: Recursive technique
  - Find the partial probabilities for each state at $t = 1$
  - Find the partial probability at the current time t from the partial probabilities at time $t - 1$.
  - Therefore recurse the partial best paths from time $t = 1$ to $t = T$ (the end of the observation sequence).
  - Find the best path by back tracking from $t = T$ to $t = 1$

10.8.2   HMM Decoding: Partial probabilities (δ's)

Initial partial probabilities δ's at time $t = 1$ are calculated using:

$$\delta_1(j) = \pi(j)b_{k_1 j}. \tag{10.6}$$

Partial probabilities δ's are calculated iteratively at time $t > 1$ using:

$$\delta_t(j) = \max_i(\delta_{t-1}(i)a_{ij}b_{k_t j}). \tag{10.7}$$

The iterative application of (10.7) is shown in Figure 10.9 (for the example of three hidden states).

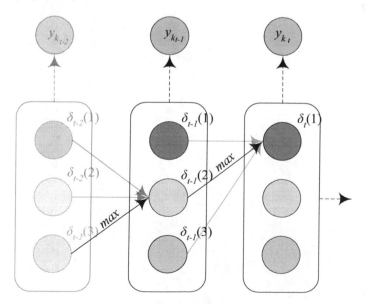

FIGURE 10.9: Iterative calculation of partial probabilities δ's using (10.7).

The path choices for each state j (given by the function $\psi_t(j)$) at each time step t need to be recorded:

$$\psi_t(j) = argmax_i(\delta_{t-1}(i)a_{ij}). \tag{10.8}$$

10.8.3   Viterbi Algorithm: Finding the final best path

Once the partial probabilities are calculated, the final best fitting sequence of hidden states is calculated using the Viterbi back tracing algorithm.

- Find the most likely state q_T^* at the end of the sequence $t = T$

$$q_T^* = \text{argmax}_j(\delta_T(j))$$

- Trace back through all the states from $t = T$ to $t = 1$

$$q_t^* = \psi_{t+1}(q_{t+1}^*)$$

- On completion, the final best path is the states corresponding to $q_1^* \ldots q_T^*$

This Viterbi recursion method from the last time step to the first is illustrated for a three hidden state HMM shown in Figure 10.10.

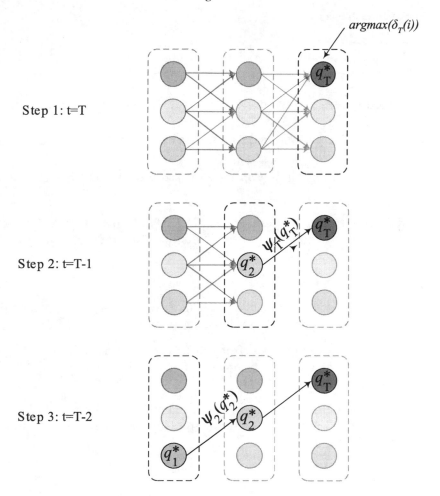

FIGURE 10.10: The Viterbi backwards iteration method to obtain the most likely sequence for a three hidden state HMM.

10.8.4 Viterbi Algorithm: Example

Figure 10.11 shows an example of the Viterbi algorithm used for the same HMM defined in Figure 10.8. This figure shows the partial probabilities δ calculated for each hidden state for two time steps. These two time steps are associated with an observation sequence medium, medium. The first column (on the left) is calculated using (10.6). The second column is calculated using (10.7). The final most likely state (with a probability of 0.1225) for the right-hand column is therefore a. Tracing the decision back gives the most likely sequence of a, a.

Partial Probabilities δ

⟶

	high	high
a	$0.5 \times 0.7 = 0.35$	Max (**0.35×0.5×0.7**, 0.06×0.2×0.7, 0.09×0.3×0.7) =0.1225
e	$0.3 \times 0.2 = 0.06$	Max (**0.35×0.25×0.3**, 0.06×0.6×0.3, 0.09×0.15×0.3) =0.02625
ə	$0.3 \times 0.3 = 0.09$	Max (**0.35×0.25×0.3**, 0.06×0.2×0.3, 0.09×0.55×0.3) =0.03675

FIGURE 10.11: Example Viterbi algorithm for HMM decoding: Using Equation (10.6) for the left-hand column and Equation (10.7) for right-hand column.

10.8.5   Viterbi Algorithm: Summary

- By backtracking rather than forward tracking, we can avoid garbled portions of the sequence for speech recognition systems.

- The Viterbi algorithm differs from the forward algorithm by replacing the summation function with a max function to calculate the partial probabilities (δ's and α's).

- Additionally, the Viterbi algorithm remembers the best route between states from the previous position to the current position via the `argmax` calculation of the ψ's.

- Final state probability can be used for evaluation.

10.9   HMMs: Learning

Both the evaluation and decoding problems assume that the HMM is already defined (through the definition of the three items $\lambda=\{\Pi, \mathbf{A}, \mathbf{B}\}$). Within the context of speech recognition, there is usually a single HMM trained to represent a phone, n-gram, [2] word or combination of words and phones. In order to train each HMM, multiple examples of the speech element to be recognised are required.

- The parameters of an HMM are often not directly measurable.

---

[2]n-grams are combinations of phones.

- Therefore there must be a learning mechanism

    - The learning problem

- The forward backward algorithm

    - This algorithm makes an initial guess of the parameters and they are iteratively updated until a termination condition (giving the final approximation).
    - The next guess at each iteration is based on a form of gradient descent derived from an error measure. This method is similar to Expectation Maximisation (EM).

10.10 Example MATLAB HMM Code

The listing below illustrates an example piece of MATLAB HMM code. This code (and output) illustrates the use of the MATLAB commands hmmgenerate, hmmdecode and hmmviterbi. These three commands use the definition of just the state transition matrix A and the emission probability matrix B. The initial state of the sequence is assumed to be in hidden state 1. Therefore the initial state probability vector is not needed (or assumed to be [1,0,0,0...]). hmmgenerate generates an observation sequence (of length 5 in this case), hmmdecode generates the partial probabilities of each hidden state at each time step and hmmviterbi generates the most likely hidden state sequence.

```
>> A = [0.65,0.35;
0.20,0.80];
>> B = [0.2 0.2 0.2 0.2 0.1 0.1 ;
0.1 0.2 0.1 0.1 0.1 0.4];
>> [seq,states] = hmmgenerate(5,A,B)
seq =
     6 5 5 6 5
states =
     1 1 2 2 2
>> hmmdecode(seq,A,B)
ans =
    0.2968 0.2958 0.2485 0.1230 0.2554
    0.7032 0.7042 0.7515 0.8770 0.7446
>> hmmviterbi(seq,A,B)
ans =
     2 2 2 2 2
```

Other MATLAB HMM functions for training such as hmmestimate and hmmtrain are also available.

10.11   Hidden Markov Models for Speech

The HMMs described so far are for general use, i.e., every hidden state can be followed by itself or any other state. A more common HMM model for speech recognition is illustrated in Figure 10.12. These temporal type HMMs have the following features:

- They are typically left to right HMMs (sequence constraint).

- There are self loops for time dilation in the HMM model (for the modelling of timing variations).

- The progression of states can only be sequential (or self-loops).

- A hierarchy of these types of models can be used to handle different speech timings, pronunciation, grammar and hidden states all within the same structure.

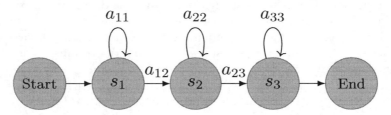

FIGURE 10.12: Typical HMM structure for ASR, e.g., where the hidden states are phones.

10.12   HMM Emission Probability Estimation for ASR

In the above examples, the observation states are discrete. Observations are often continuous, e.g., the original barometric readings. A method of modelling continuous readings is therefore required.

- Discrete Probabilities

    - Vector Quantisation (VQ)

    - In order to derive observation states the input feature vectors are quantised into VQ bins.

    - e.g., for barometric reading

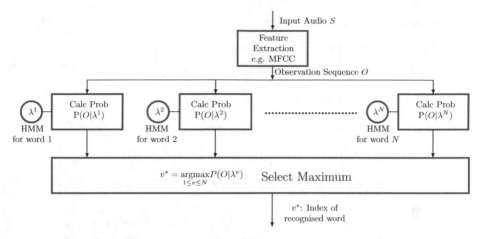

FIGURE 10.13: HMMs for single-word recognition.

- High is pressure > 110
- Medium is pressure $90 - 110$
- Low is pressure < 90

• Gaussian Densities

  - Parametrically controlled density function defined by mean and co-variance.

• Gaussian Mixtures

  - Weighted combination of Gaussians (see GMMs below)

• Neural Networks (see section 10.16).

10.13   Single-Word Speech Recognition System

Single-word recognition systems are an alternative and more constrained ASR system relative to continuous recognition systems. They have short pauses between spoken words and are primarily used in small vocabulary command applications such as name-dialling. Figure 10.13 shows an HMM based single-word speech recognition system. A single model (in this case an HMM) is defined per word and the most likely posterior probability for the range of word models can be considered to be the recognised word. Deep network keyword recognition systems are described in more detail in section 10.16.

10.14   HMMs for ASR: Summary

- Well-defined mathematical basis and structure.

- Errors in the analysis do not propagate and accumulate.

- Does not require the prior temporal segmentation of the signal.

- Temporal property of speech is accounted for.

- Used with features usually MFCC or MFCC Δs or $\Delta\Delta$s (or similar invariant features).

- For individual word recognition an HMM can be formed for each word and a maximum likelihood method can be used to choose the found word.

- Can be combined with a lexical and grammatical model to perform continuous speech recognition.

- Can be combined with GMM or DNN emission probability estimation.

10.15   Emission Probability Estimation using Gaussian Mixture Models (GMMs)

Before the advent of Deep Neural Networks for ASR combined HMM-GMM systems gave the state of the art performance for speech recognition systems. GMM models are able to flexibly model the emission probabilities of continuous MFCC type features extracted from the input audio. A GMM is defined as a weighted sum of Gaussian PDFs:

$$p(\boldsymbol{x}) = \sum_{i=1}^{K} w_i \mathcal{N}(\boldsymbol{x}|\boldsymbol{\mu}_i, \boldsymbol{\Sigma}_i) \tag{10.9}$$

$$\sum_{i=1}^{K} w_i = 1, \tag{10.10}$$

where; $p(\boldsymbol{x})$ is the GMM probability; $\boldsymbol{x}$ is observation feature vector (defined in Equation (2.63) (e.g., features from an MFCC type of extraction process from the input audio); K is the number of Gaussian distributions in the model; $\boldsymbol{\mu}_i$ and $\boldsymbol{\Sigma}_i$ define the mean vector and covariance of the ith model, respectively. As indicated, the weights w_i must sum to unity. An HMM-GMM model is therefore defined as the state transition matrix A, the initial state probability vector Π and the parameters of the GMM ($\{w_i, \mu_i, \Sigma_i\}$ for $i= 1,2,\ldots,K$). All of these parameters can be learnt using an iterative EM type method. The general structure of an HMM-GMM system for ASR is shown in Figure 10.14.

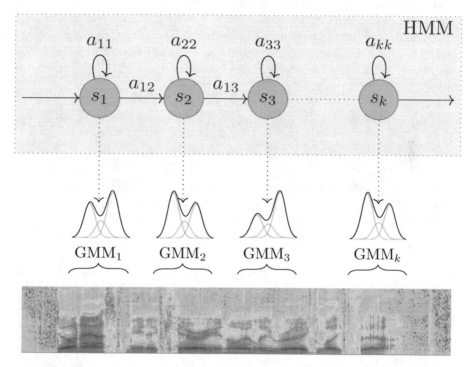

FIGURE 10.14: ASR recognition system using HMMs and GMMs.

10.16   ASR using Deep Neural Networks (DNNs)

10.16.1   Deep Neural Networks

As described in chapter 2, machine learning is a generic term for the automatic discrimination of input patterns into defined classes.

In recent years excellent generic classification performance has been achieved with Deep Neural Networks (DNNs). This is due in part to the recent developments of newly formed activation functions, generalisation and regularisation methods but mainly due to the large amounts of data available to train such systems and the associated computational power that is necessary and now available (in the form of very high powered GPU cards). Furthermore, the densely connected, hierarchical and deep nature of DNNs are not only able to classify but generate characterising features of progressive complexity from the input to the output layers.

DNNs are artificial neural networks that use a feed-forward structure and a large number of hidden units often densely connected between the input nodes and the output nodes. Each hidden unit j or "neuron" uses an "activation function" (denoted as ϕ) to map the total input from the layer below $\{x_1, x_2, \ldots x_n\}$, to its scalar output y_j. This scalar output y_j is then the input sent to the upper layer above and so on. The workings of a single neuron can be summarised as follows:

$$v_j = b_j + \sum_{i=1\ldots n} x_i w_{ij} \tag{10.11}$$

$$y_j = \phi(v_j), \tag{10.12}$$

where ϕ is the activation function, i is an index over lower units, b_j is the biasing offset for this unit j, and w_{ij} is the weight of the connection from unit j to unit i in the layer below. This is illustrated in Figure 10.15.

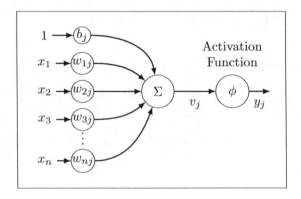

FIGURE 10.15: Single neuron.

Typical activation functions include **tanh** (the hyperbolic tangent that compresses any scalar input into the range -1 to 1) or **ReLU** (Rectified Linear Unit). Such functions are chosen often because their gradients can be easily calculated as this is key when training the network.

The output units within a multiclass classification DNN often converts its total input, $\{x_1, x_2, \ldots x_n\}$, into a class probability, p_j, by using the well-known **softmax** function.

$$p_j = \frac{exp(x_j)}{\sum_k exp(x_k))} \tag{10.13}$$

A multilayered neural network is illustrated in Figure 10.16.

The weights of all neuron units are initiated within training with random

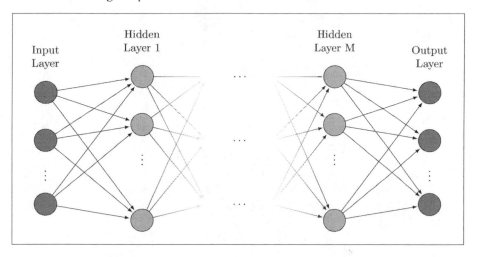

FIGURE 10.16: Multilayered neural network.

or pseudo-random values. Weights are then iteratively updated using methods such as a batch based back propagation method (Stochastic Gradient Descent) using the ground truth outputs and a training dataset. Many techniques are often used to control a network's tendency to over or underfitting. Methods to prevent overfitting and promote generalisation are often key in producing successful classification networks such as those used for speech recognition. These methods include techniques such as L_1, L_2 and dropout based regularisation [7].

10.16.2   DNNs for ASR

Initially, DNNs were used in conjunction with HMMs as a direct replacement of GMMs to model HMM emission probabilities, i.e., they modelled the association between phones and acoustic features such as MFCC features. These systems were initially based on Restricted Boltzmann Machines (RBMs) to model the input data using undirected probabilistic models to generate Deep Bayesian Networks (DBNs). These systems were able to dramatically improve the classification performance of ASR. Specifically the TIMIT dataset error rate for the state of the art GMM-HMM system was approximately 26% and this dropped to 20.7% in 2009 through the use of combined DNN-HMM systems [7, 11, 12]. An excellent overview of a selection of DNN-HMM architectures from this era is given by Hinton et al. [6].

Figure 10.17 shows how a DNN can be used to interface between an audio input and an HMM for speech recognition in a similar way to that described by Hinton et al. [6] and Dahl et al. [5].

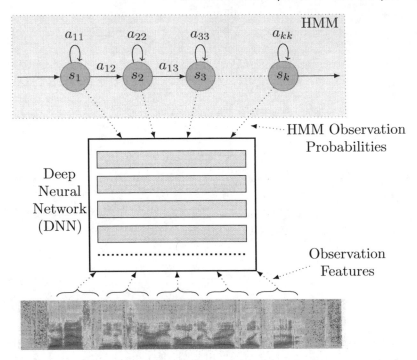

FIGURE 10.17: Example structure of an DNN-HMM for ASR.

10.17 Moving Forward: HMM-Free ASR

Although the emphasis within this chapter has been based on the structure of HMMs and different systems (e.g., GMMs and DNNs) to estimate the HMM emission probabilities there are currently emerging "end-to-end" DNNs that are no longer based on HMMs.

10.17.1 Recurrent Neural Networks (RNNs)

Figure 10.18 illustrates the structure of a Recurrent Neural Network (RNN), i.e., there is no "phone model" or feature extraction stage: everything is learnt from a very large dataset. Such systems commonly do not use any explicit feature extraction as feature representations are learnt from the input training data that is most key for classification rather than hand defined features. This figure shows how temporal connections (or "edges") are made between the neighbouring temporal versions of the networks.

10.17.2 Long Short-Term Memory (LSTM) Systems

Conventional RNNs just combine previously predicted neural outputs of the previous temporal timestep with the current neural inputs. LSTM methods add a local state to each network element that is updated flexibly. Due to this memory, an arbitrarily long "context window" can be retained to optimise speech recognition without the need for HMMs. It is believed that LSTM based RNNs often outperform HMM methods as HMM are only able to model a modest number of hidden states. Furthermore, the use of LSTM methods removes the need to use hierarchical models such as HMMs to characterise an arbitrarily long context for recognition. LSTM methods are also able to directly use the output from an STFT and learn characterising representations rather than handcrafted features (as described in the previous chapters).

Each element of a LSTM neural network is termed a "memory cell" and is made up of the following components:

Internal State: An internal state (or memory) is maintained within the "memory cell".

Input Node: The input node takes the current input and the previous prediction as the input to a specifically defined activation function (typically a tanh) function.

Input Gate: The input gate takes the same input as the input node and commonly outputs a number in the range from 0 to 1. The output is multiplied by the output of the input node and therefore "gates" the input's elements according to the weights of the input gate.

Forget Gate: The forget gate also has the same input as both the input node and gate and provides a mechanism to flush the unneeded elements from the internal state.

Output Gate: The output of the entire memory cell is the multiplication of the memory cell state and the output gate.

Such a single "memory cell" connected forwards (and possibly backwards) temporally can form any or all the RNN network elements (depending on the architecture) illustrated in Figure 10.18. An illustration of how previous outputs and cell state are passed temporally is shown in Figure 10.19. This figure shows a single "memory cell" unit that could be included within a general RNN structure shown in Figure 10.18. A more in-depth definition of RNN-LSTM "memory cells" can be found in the tutorial by Lipton and Berkowitz [10] which also describes some of the more recent advances such as "peephole" and "attention" variants.

An example of such a system is the "Deep Speech" system [1]. This system combines an RNN with an N-gram language model to achieve (at the time of its publishing) the best classification performance on the Hub500 dataset (an error rate of 16.0%). The development of RNNs for speech recognition was initiated by Graves et al. [8] who used them with LSTM methods to achieve (at the time) the state of the art TIMIT error rate.

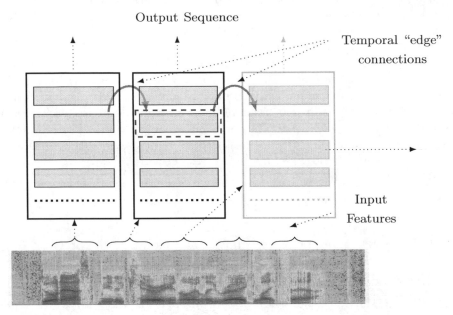

FIGURE 10.18: Example structure of an RNN for ASR. Each node (e.g., the one in the dashed box) can be implemented using a LSTM "memory cell" as shown in Figure 10.19.

Many current individual word (or keyword) identification systems use Convolutional Neural Networks (CNNs) more similar to those used for image recognition. These methods just use a 2D spatial-temporal representation of the input audio as the input to the training and testing of the system. Such a keyword recognition system has been developed by Microsoft (Xiong et al. [14]) and is available as a Tensorflow model. CNNs have also been applied directly to ASR by Abdel et al. [2].

10.18   HMM, GMMs and DNNs: Summary

- HMMs are able to model the timing variation within speech recognition.

- GMMs are used within HMM-GMM systems to flexibly model the emission probabilities of an HMM.

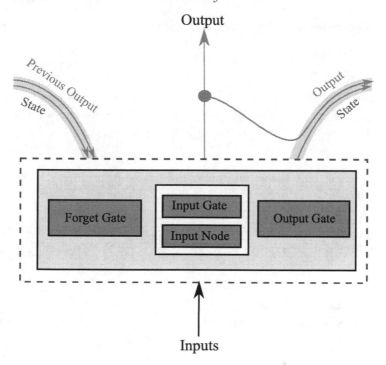

FIGURE 10.19: Single "memory cell" of a LSTM RNN network (such as Figure 10.18).

- Deep Neural Networks (DNNs) used to model the emission probabilities of HMMs provided a large increase in the performance of ASR systems.

- End-to-end systems (e.g., using RNNs) now provide state of the art speech recognition performance. However, the structure represented by HMMs remains an important tool in speech analysis.

- LSTM system combined with RNNs are able to improve the performance of HMM ASR system. This is due to the ability of LSTM systems to learn features and hold an internal state that contextualises the recognition process from an arbitrarily long temporal window.

- Although end-to-end RNN based ASR systems now provide many of the state of the art systems, HMM-DNN based systems are still used in contemporary systems: The "hey-Siri" voice trigger algorithm developed by Apple on IPhones still uses an HMM-DNN based system (as of 2018).

10.19 Exercises

Exercise 10.1
Repeat the example shown in figure 10.8 but with the observation sequence {high, high, high}, i.e., find the partial probabilities for the last column and the total probability of the whole sequence.

Exercise 10.2
Repeat the example shown in figure 10.8 but with the observation sequence {high, high, medium}, i.e., find the partial probabilies for the last column and the total probability of the whole sequence.

Exercise 10.3
What are the advantages of using a RNN-LSTM over an HMM based system?

Bibliography

[1] A. Hannun, et al. Deep speech: Scaling up end-to-end speech recognition. arxiv preprint arxiv:1412.5567. 2014.

[2] O. Abdel-Hamid, A-R Mohamed, H. Jiang, and G. Penn. Applying convolutional neural networks concepts to hybrid nn-hmm model for speech recognition. In *Acoustics, Speech and Signal Processing (ICASSP), 2012 IEEE International Conference on*, pages 4277–4280. IEEE, 2012.

[3] J. Baker. The Dragon system–An overview. *IEEE Transactions on Acoustics, Speech, and Signal Processing*, 23(1):24–29, 1975.

[4] L.E. Baum and T. Petrie. Statistical inference for probabilistic functions of finite state markov chains. *The Annals of Mathematical Statistics*, 37(6):1554–1563, 1966.

[5] G.E. Dahl, D. Yu, L. Deng, and A. Acero. Context-dependent pre-trained deep neural networks for large-vocabulary speech recognition. *IEEE Transactions on Audio, Speech, and Language Processing*, 20(1):30–42, 2012.

[6] G. Hinton, et al. Deep neural networks for acoustic modeling in speech recognition: The shared views of four research groups. *IEEE Signal Processing Magazine*, 29(6):82–97, 2012.

[7] I. Goodfellow, Y. Bengio, A. Courville, and Y. Bengio. *Deep learning*, volume 1. MIT press Cambridge, 2016.

[8] A. Graves, A.-R Mohamed, and G. Hinton. Speech recognition with deep recurrent neural networks. In *Acoustics, Speech and Signal Processing (ICASSP), 2013 IEEE International Conference on*, pages 6645–6649. IEEE, 2013.

[9] F. Jelinek, L. Bahl, and R. Mercer. Design of a linguistic statistical decoder for the recognition of continuous speech. *IEEE Transactions on Information Theory*, 21(3):250–256, 1975.

[10] Z.C. Lipton, J. Berkowitz, and C. Elkan. A critical review of recurrent neural networks for sequence learning. arxiv preprint arxiv:1506.00019. 2015.

[11] A-R Mohamed, G. Dahl, and G. Hinton. Deep belief networks for phone recognition. In *Nips Workshop on Deep Learning for Speech Recognition and Related Applications*, volume 1, page 39. Vancouver, Canada, 2009.

[12] A-R Mohamed, G.E. Dahl, and G. Hinton. Acoustic modeling using deep belief networks. *IEEE Transactions on Audio, Speech, and Language Processing*, 20(1):14–22, 2012.

[13] L.R. Rabiner. A tutorial on hidden markov models and selected applications in speech recognition. *Proceedings of the IEEE*, 77(2):257–286, 1989.

[14] W. Xiong, et al. The microsoft 2016 conversational speech recognition system. In *Acoustics, Speech and Signal Processing (ICASSP), 2017 IEEE International Conference on*, pages 5255–5259. IEEE, 2017.

11

Speech Coding

CONTENTS

> In making a speech one must
> study three points: first, the
> means of producing persuasion;
> second, the language; third the
> proper arrangement of the
> various parts of the speech.
>
> Aristotle

11.1   Speech Coding: Introduction

Previously in Chapter 7 coding focused on general and wideband audio codecs such as MPEG1 audio layer3 (MP3). This chapter focuses on codecs specifically designed for speech coding. These codecs have a long history of standardisation and have in a general sense improved in quality (for equivalent bandwidth) over this time. Speech coding techniques are specifically designed for speech signals. They are therefore suboptimal for music and other general audio applications, i.e., using a mobile phone to listen to music over a GSM line is not a good idea.

The human speech signal has the following characteristics:

- Usually sampled at 8kHz

- Band limited between 200 and 3400Hz, i.e., this band contains all "speech information"

There are three main types of speech codec

- Waveform Codecs

  - Usually high bit rates/high quality

- Speech Vocoders

  - Can be very low bit rate but can sound synthetic

- Hybrid Codecs

  - Use techniques from both waveform codecs and speech vocoders and give a good compromise in quality and bitrate

Figure 11.1 shows a comparison of the ranges of rate-quality for the three different types of speech codec.

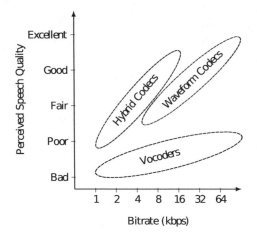

FIGURE 11.1: Rate-quality graphs for the three main speech codec types. N.B. The y axis is indicative rather than exact mean opinion scores.

11.2   Waveform Codecs

Waveform codecs are simple codecs and not always specifically designed for speech. They are often of high quality approximately above 16kbps; quality significantly deteriorates below this threshold. The three main types of waveform codecs are (in increasing levels of quality) Pulse Code Modulation (PCM), Differential Pulse Code Modulation (DPCM) and Adaptive Pulse Code Modulation (ADPCM). 8kHz, 16 bit sampling is the most commonly found version of speech based PCM, DPCM and ADPCM and the reported bitrates below are based on this assumption.

11.2.1 Pulse Code Modulation (PCM)

PCM is often considered the "raw" digital format, being a sampled and linearly quantised representation of the signal. It is the coding format of the .wav and .aiff file types commonly used to store raw audio. PCM has the following characteristics.

- For linear quantisation, PCM has a bit rate of 128kbps (mono, 16 bits per sample, sampled at 8kHz).

- Logarithmic quantisation can lead to a high quality 64kbps bitrate stream (known as companding).

    - u-law and A-law are logarithmic quantisation standards in the U.S. and Europe, respectively.

- u-law companded PCM has been standardised as G.711 at 64kbps.

11.2.2 Differential Pulse Code Modulation (DPCM)

DPCM is similar to PCM except that the difference between the current sample and a prediction based on its predecessors is coded, thus reducing entropy. DPCM can code to the same quality of 64kbps PCM at 48kbps.

11.2.3 Adaptive Pulse Code Modulation (ADPCM)

ADPCM is similar to DPCM except that the samples are adaptively quantised according to the signal; ADPCM at 32kbps has a similar perceptual quality to 64kbps PCM. 16, 24, and 40kbps versions have also been standardised.

11.3 Speech Vocoders

Figure 11.2 shows the structure of a traditional channel vocoder typical of those developed in 1940s, 1950s and 1960s [1]. The signal is analysed by a set of bandpass filters and the results are encoded. Within this system the excitation signal is either:

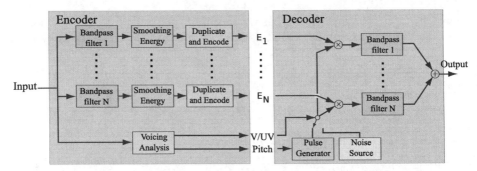

FIGURE 11.2: Historical Vocoder encoding structure.

- Signalled as either unvoiced or voiced. If the speech (frame) is identified as being voiced then the pitch is also obtained and coded. At the decoder, the excitation is recreated through a pulse generator (or similar) for a voiced source or a noise generator for an unvoiced source.

- A spectrally flattened version of the original baseband signal.

Vocoders are intelligible but can sound very synthetic.

11.3.1   LPC Vocoders

LPC vocoders are very similar to filterbank vocoders. The spectral envelope is estimated using an LPC analysis (examples of LPC analysis is described in Chapter 9).

As with most speech processing, the analysis is implemented on short temporal "frames" of the signal. Filter parameters are updated every 15 to 30ms. The general structure of a pure LPC encoder is shown in Figure 11.3 [1].

Once the spectral envelope has been estimated using LPC analysis, the excitation signal can be represented using one of a number of methods. These methods include the two described above. Two other common early methods of excitation signal coding for LPC vocoders are VELP and RELP.

- **VELP**: Voice-Excited Linear Prediction

  - LPC analysis used to model the spectral envelope. Excitation signal downsampled and quantised.

- **RELP**: Residual-Excited Linear Prediction

  - LPC analysis also used to model the spectral envelope. Excitation created through downsampled and quantised version of a baseband version of the speech signal.

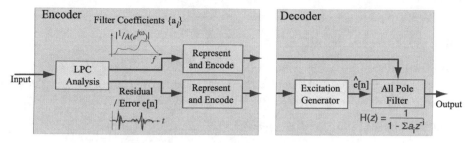

FIGURE 11.3: LPC Vocoder structure.

11.4 Hybrid Codecs: Analysis by Synthesis (AbS)

Hybrid codecs provide a combination of waveform and vocoder methods. In older speech vocoders, the excitation signal is modelled as either a voiced or unvoiced synthetic signal. This leads to a synthetic speech result. In hybrid codecs, the excitation signal $e[n]$ is modelled by trying to most closely match the reconstructed signal $\hat{s}[n]$ to the original signal $s[n]$ (when the excitation signal is filtered by the LPC filter). This is most often implemented in the time domain using an Analysis by Synthesis (AbS) technique illustrated by the iterative coding block diagram shown in Figure 11.4.

AbS techniques divide the signal into frames of typically 20ms. For each frame, parameters are determined for a synthesis LPC filter, and then the excitation signal of this filter is determined. An excitation signal is encoded which, when filtered by the LPC synthesis filter, minimises an error between the reconstructed speech and the input speech. The synthesis $H(z)$ filter is typically encoded using an all-pole form, i.e., LPC (or converted to LSP coefficients):

$$H(z) = \frac{1}{A(z)} = \frac{1}{\left(1 - \sum_{j=1}^{P} a_j z^{-j}\right)}, \tag{11.1}$$

where $\{a_j\}$ are the coefficients and P is the filter order (P is typically 10: 8 for GSM). AbS methods are usually distinguished by the method of encoding the excitation signal. These methods include MPE, RPE and CELP:

Multi Pulse Excited (MPE)

- Excitation signal is given by a fixed number of non-zero pulses for every frame of speech.

- The positions and amplitudes for these non-zero pulses are transmitted to the decoder.

- Typically 4 pulses per 5 ms are used.

Regular Pulse Excited (RPE)

- The pulses are regularly spaced at a fixed interval. Only the position and amplitude of the first pulse is required to be transmitted, thus leading to a decrease in bits compared to MPE.

Code Excited Linear Prediction (CELP)

- Excitation signal quantised using vector quantisation.

- Therefore a codebook of excitation signals is required.

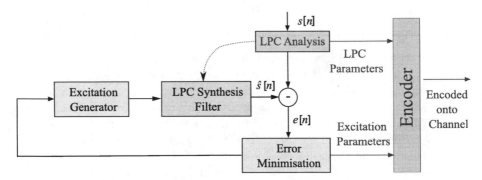

FIGURE 11.4: Analysis by synthesis LPC codec structure: $s[n]$ is the input signal, $\hat{s}[n]$ is the estimated signal, $e[n]$ is the error (difference between the estimated signal $\hat{s}[n]$ and the actual input signal $s[n]$).

11.4.1 Analysis by Synthesis Filters

There are two types of filters within AbS techniques: pitch filters and weighting filters.

Pitch Filters

A pitch filter can be applied to both Multi-Pulse Excited (MPE) and Regular Pulse Excited (RPE) codecs. A pitch filter is always applied to CELP codecs. It is either implemented directly or using an adaptive codebook. The pitch filter is included to model the long-term periodicities present in voiced speech. The addition of the pitch synthesis filter means that only the difference between the excitation signal and a scaled version of what it previously was, has to be transmitted.

Perceptual Weighting Filters

Error weighting filters are used to modify the spectrum of the error to reduce the subjective loudness of the error. The error signal in frequency ranges where the speech has high-energy will be at least partially masked by the speech. The weighting filter emphasises the noise within the frequency ranges where the speech content is low (or at least not perceptually important for speech perception).

11.5 LPC Parameters for Speech Encoding

LPC vocoders require the transmission of the LPC parameters $\{a_j\}$ within the all-pole filter representation (as illustrated in Figure 11.3). It has been found that the full parameter set $\{a_j\}$ is not good for transmission of speech data. This is because these parameters are very susceptible to quantisation noise. Furthermore, the parameters with higher values of j are much more susceptible to quantisation. Finally, they are not easily interpolated from one frame to the next. LPC parameters are therefore often coded in an alternative form. A typical alternative form is known as Line Spectral Pairs (LSPs). LSPs have the following characteristics:

- The roots of the original LPC filter transfer function $A(z)$ can lie anywhere within the unit circle.

- The transfer function $A(z)$ is separated into two transfer functions $P(z)$ and $Q(z)$ which have roots on the unit circle in pairs.

$P(z)$ and $Q(z)$ can be obtained by summing the filter coefficients with their time reverse (given the definition in (11.1)):

$$P(z) = A(z) - z^{-(P+1)}A(z^{-1}), \tag{11.2}$$
$$Q(z) = A(z) + z^{-(P+1)}A(z^{-1}), \tag{11.3}$$

where $P(z)$ and $Q(z)$ are the even and odd-symmetric filters, respectively whose roots lie on the unit circle. These equations are implemented in MATLAB Listing 11.1 (lines 10 and 11).

Figure 11.5 shows a screen shot of the MATLAB LSP roots GUI.[1] This shows an application that is able convert LPC coefficients extracted from a speech frame into two polynomials $P(z)$ and $Q(z)$ whose roots are all on the unit circle.

MATLAB Listing 11.1 shows a MATLAB LPC analysis of a segment of audio followed by the conversion of the LPC coefficients to LSP coefficients. The poles of both the LPC and LSP coefficients are shown in Figure 11.6. This shows that the LSP coefficients lie on the unit circle and are placed close to the positions of the poles of the LPC analysis.

---

[1]Available at https://uk.mathworks.com/matlabcentral/fileexchange/45436-lsp-roots.

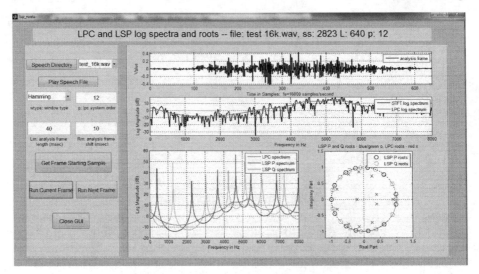

FIGURE 11.5: MATLAB LSP roots GUI.

Listing 11.1: LSP Test Code

```
1   [x,fs]=audioread('paulhill2.wav',[24020 25930]);
2   x=resample(x,10000,fs); fs=10000;
3   t=(0:length(x)-1)/fs;
4   x1 = x.*hamming(length(x));
5   ncoeff=10; % rule of thumb for formant estimation
6   a=lpc(x1,ncoeff);
7   zp = zplane([],a); hold on;
8
9   %Convert to LSP coefficients (p and q).
10  p=[a 0]+[0 fliplr(a)]; %P(z) = A(z)-z^{-(P+1)}A(z^{-1})
11  q=[a 0]-[0 fliplr(a)]; %Q(z) = A(z)+z^{-(P+1)}A(z^{-1})
12
13  [zP,pP,kP] = tf2zp(1,p);
14  [zQ,pQ,kQ] = tf2zp(1,q);
15
16  plot(pP,'r+');
17  plot(pQ,'g+');
```

MATLAB Listing 11.2 shows a MATLAB LPC analysis of a segment of audio followed by the conversion of the LPC coefficients to Line Spectral Frequencies (using the MATLAB function poly2lsf). These frequencies are then plotted around the unit circle and contain both the zero positions of polynomials P and Q extracted above using LSP conversion (see Figure 11.7).

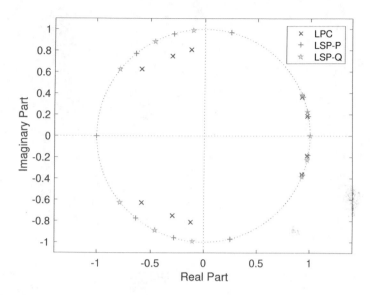

FIGURE 11.6: Position of LPC poles and their converted LSP poles given in MATLAB Listing 11.1.

Listing 11.2: LSF Test Code

```
1  [x,fs]=audioread('paulhill2.wav',[24020 25930]);
2  x=resample(x,10000,fs); fs=10000;
3  t=(0:length(x)-1)/fs;
4  x1 = x.*hamming(length(x));
5  ncoeff=10; % rule of thumb for formant estimation
6  a=lpc(x1,ncoeff);
7  zp = zplane([],a); hold on;
8
9  %Convert to LSF frequencies
10 LSFcoef = poly2lsf( a ) ;
11 LSFpt = [ exp(LSFcoef*i) ; conj(exp( LSFcoef*i ) ) ] ;
12 plot(LSFpt,'g^');
```

11.6 Speech Coding Standards

Table 11.2 shows a list of some of the most commonly found speech codecs. This table shows the standard name, the coding method used, bitrate and the experimental Mean Opinion Score (MOS). The MOS is derived from listen-

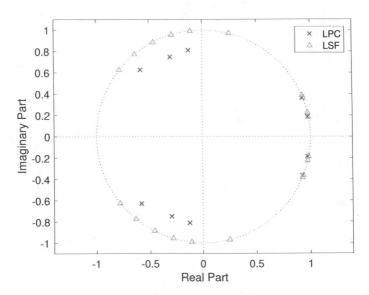

FIGURE 11.7: Position of LPC poles and their converted LSF poles (around the unit circle) given in MATLAB code 11.2.

ing experiments averaged over a large number of subjects. Table 11.1 shows a table of MOS scores valued from 1 to 5 (increasing in perceptual quality). Figure 11.8 shows an illustration of these MOS values for a selection of speech codecs (taken from the result given in Table 11.2). More up to date evaluation methods have been used recently in speech coding. Figure 11.9 shows a graph taken from the Opus website and illustrates the Rate-Quality graphs of a number of speech and wideband audio codecs. It can be seen that Opus is very effective at coding speech and audio across a very wide range of bitrates.

There are several names not otherwise defined in table 11.2. They are:

ACELP Algebraic CELP: An algebraic codebook is used to index excitation pulses.

CS-ACELP Conjugate Structure Algebraic CELP.

RPE-LTP Is a Regular Pulse Excited (RPE) based codec combined with Long Term Prediction (LTP).

LD-CELP Is a Low Delay CELP codec.

GSM-FR Groupe Spécial Mobile: Full Rate. Often just known as GSM. This is the most common cellular speech codec.

GSM-EFR Enhanced Full Rate GSM.

GSM-HR Half-Rate GSM.

TABLE 11.1: Mean Opinion Score (MOS) Metric

MOS	Quality	Impairment
5	Excellent	Imperceptible
4	Good	Perceptible, but not annoying
3	Fair	Slightly annoying
2	Poor	Annoying
1	Bad	Very annoying

TABLE 11.2: Speech Coding Standards

Codec / Method	Year	Bitrate: kbps	MOS
G.711: PCM	1972	64	4.1
FS1015: LPC-10	1984	2.4	2.24
GSM-FR: RPE-LTP	1989	13	3.6
FS1016: CELP	1991	4.8	3
G.726: ADPCM	1991	24	3.5
G.726: ADPCM	1991	32	4.1
G.728: LD-CELP	1994	16	3.61
G.723.1: ACELP	1995	30	3.6
GSM-EFR: ACELP	1995	12.2	4.1
G.729: CS-ACELP	1995	8	3.92
GSM-HR: VSELP		5.6	3.5
Speex (NB): CELP	2003	2.15-44	2.8-4.2
Opus: SILK/CELT	2012	6-510	

MOS data taken from [2].

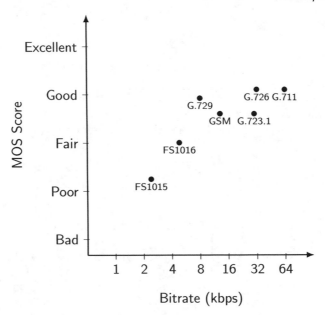

FIGURE 11.8: Mean Opinion Scores (MOS) of a selection of standardised speech codecs. From data taken from [2] (i.e., Table 11.2).

11.7 Comparison of Speech Coding Standards

A selected number of speech codecs are now described in more detail.

11.7.1 G.711 [3]

G.711 is a waveform codec (standardised in 1972) that uses PCM and u-law companding to generate very good MOS scores (usually over 4) at a bitrate of 64kbps.

11.7.2 FS1015

FS1015 is a vocoder based coder (standardised in 1984) using what is known as the LPC-10 method at a rate of 2.4 kbps. It encodes a simple voiced/unvoiced flag together with the pitch of vibration if voiced.

- Frame size is 20ms. 50 frames/sec. 2400 bps = 48 bits/frame.

- 10 LSP parameters per frame to define the filter.

- Extremely low bit rate.

- Military applications: Encryption.

- A breakdown of parameters given in table 11.3.

TABLE 11.3: FS1015: LPC-10 Speech Codec

Parameters	Number of Bits
LPC(LSP)	34
Gain	7
Voiced/Unvoiced and Period	7
Total	**48**

11.7.3 FS1016

FS1016 is a CELP based coded (standardised in 1991) at a rate of 4.8 kbps.

- Frame size is 30ms. 33.3 frames/sec. 4800 bps = 144 bits/frame.

- Each frame divided into 4 subframes. The codebook for vector quantisation contains 512 vectors for each subframe.

- The gain is quantized using 5 bits per subframe.

- A breakdown of parameters given in table 11.4. Note FEC denotes Forward Error Correction; a type of redundant coding for error resilience.

TABLE 11.4: FS1016 Parameter Breakdown

Parameters	Number of Bits
LSP	34
Pitch Prediction Filter	48
Codebook Indices	36
Gains	20
Sync	1
FEC	4
Expansion	1
Total	**144**

11.7.4 GSM-FR [7]

The GSM-FR (also known as just GSM) is based on a Regular Pulse Excited (RPE) codec (standardised in 1989 for use within cellular phone networks) at a bitrate of 13kbps.

- Frame size is 20 ms. 50 frames/sec. 13000 bps = 260 bits/frame.

- 8 LPC parameters per frame to define the filter.

- These 8 parameters are Log Area Ratios (LARs) equivalent of LSPs.

- Frame subdivided into 4 subframes of 5ms each.

- Includes a Long Time Prediction (LTP) filter (equivalent of the pitch prediction filter).

- A breakdown of parameters given in table 11.5.

TABLE 11.5: GSM Codec

Parameters	Number of Parameters	Number of Bits
LARs	8 per frame	36
LTP lag	1 per subframe (7 bits)	28
LTP gain	1 per subframe (2 bits)	2
RPE grid position	1 per subframe (2 bits)	8
Block amplitude	1 per subframe (6 bits)	24
RPE Pulses	13 per subframe (3 bits)	156
Total		**260**

11.7.5   Speex [6]

Speex is a modern hybrid codec first initiated in 2002. It is a patent-free open source and free format for speech coding. Speex was specifically designed for the sampling rates of 8, 16 and 32 kHz (narrowband, wideband and ultra-wideband, respectively) and is based on Code-Excited Linear Prediction (CELP).

It targets bitrates from 2.2 to 44kbps and therefore spans a wide range of perceived qualities (and MOS scores). Furthermore, it has the facility to change bitrate adaptively in real time. It has also targeted networks that may be error-prone and therefore has an integrated error correction facility.

11.7.6   Opus [4]

Opus is a modern hybrid codec that was born out of developments by Skype and the Xiph.Org Foundation. It has been standardised by the Internet Engineering Task Force (IETF). It can support and efficiently encode general audio and speech in one format. It is also designed to be low-latency and have low-complexity. In many tests it is has shown competitive and state-of-the-art performance compared to both modern speech codecs such as Speex and high bitrate wideband audio encoders such as MP3, AAC and HE-AAC [5].

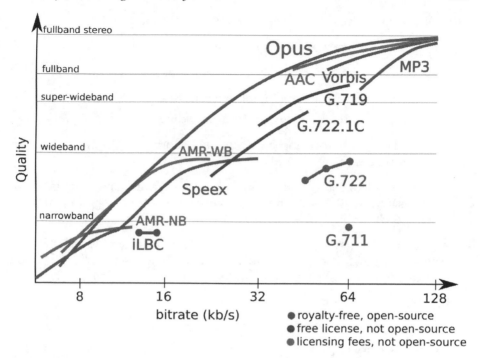

FIGURE 11.9: Rate-Quality graphs for modern speech and wideband codecs (obtained from the Opus website [5]). Copyright © Opus, licensed under CC BY 3.0 https://creativecommons.org/licenses/by/3.0/.

Opus uses and combines both an LPC based speech orientated algorithm (SILK) and a lower latency algorithm based on a MDCT transform (CELT). Opus has been designed to adjust the complexity, audio bandwidth and bitrate adaptively and seamlessly at any time while switching between (or combining) the SILT and CELT codecs. Figure 11.9 shows a rate-quality comparison of Opus with many other audio codecs (taken from the Opus website [5]) and shows the claimed state-of-the-art performance of Opus compared to other comparable codecs. The SILK codec was designed and used by Skype based on LPC methods for VoIP speech calls. The CELT codec is an acronym for "Constrained Energy Lapped Transform". It uses some CELP principles but processes the signal exclusively within the frequency domain.

11.8 Speech Coding: Summary

- There are many different techniques and standards for speech coding.

- Waveform codecs are the de-facto standard for audio coding.

    - However, they have a high bit rate.

- Speech Vocoders are the traditional method of speech coding.

    - They can produce intelligible speech at extremely low bit rates.
    - However, they are very artificial sounding.

- Hybrid coders give a compromise in rate-distortion performance between waveform codecs and vocoders.

- Hybrid coders using a type of LPC filter and excitation modelling in an analysis by synthesis method have become the most popular standards for speech coding (GSM, CELP, etc.).

- The most up-to-date speech codecs (such as Speex and Opus) are very flexible in their coding capabilities and combine many different coding methods to offer state-of-the-art rate-distortion performance.

11.9   Exercises

Exercise 11.1
Find an expression for $A(z)$ in terms of $P(z)$ and $Q(z)$ (as defined in (11.2) and (11.3)).

Exercise 11.2
Why are LSP coefficients often preferred to LPC coefficients for speech coding?

Bibliography

[1] Dan Ellis Course Notes, Columbia University: Speech and Audio Processing: ELEN E6820: http://www.ee.columbia.edu/~dpwe/e6820/outline.html: accessed 2018-01-01.

[2] S. Karapantazis and F-N Pavlidou. Voip: A comprehensive survey on a promising technology. *Computer Networks*, 53(12):2050–2090, 2009.

[3] ITU-T Recommendation G.711: Pulse Code Modulation (PCM) of Voice Frequencies: November 1988.

[4] Opus Codec: http://opus-codec.org/: accessed 2018-01-01.

[5] Opus Comparisons: http://opus-codec.org/comparison/: accessed 2018-01-01.

[6] Speex Codec: http://www.speex.org: accessed 2018-01-01.

[7] ETSI: EN 300 961 v.8.0.2: Digital Cellular Telecommunications System (Phase 2+) (GSM); Full Rate Speech; Transcoding: November 2000.

12

Musical Applications

CONTENTS

> Music expresses that which
> cannot be said and on which it is
> impossible to be silent.
>
> ---
>
> Victor Hugo

Up to this point, this book has concentrated on the synthesis and manipulation of audio for utilitarian reasons. This chapter now focuses on the synthesis and manipulation of audio for creative purposes.

12.1 Musical Synthesis: Brief History

12.1.1 The Theremin: 1920s

The Theremin was one of the first electronic instruments and was patented by Leon Theremin of Russia in 1928. It consisted (in its original form) of two metal antennas. These antennas sense the proximity of the player's hands. One antenna controls the frequency of an electronic oscillator and the other the output amplitude. The performer would move their hand along the one rod to change pitch, while simultaneously moving their other hand in proximity to the antenna. Control (and more modernly, parameterisation) of oscillator frequency and amplitude are the two key aspects of musical synthesis. Figure 12.1 shows an original Theremin being played in 1930 by Alexandra Stepanoff. The principle of modulating amplitude and frequency as exemplified by the Theremin and is key for all forms of musical synthesis.

FIGURE 12.1: Alexandra Stepanoff playing a Theramin in 1930. (Public Domain Image.)

12.1.2 Musique Concréte: 1940s

Musique Concréte is an experimental method of musical composition and synthesis using raw material usually in the form of recorded sounds (originally on discs or magnetic tape). This method of composition and music production was pioneered by the French composer Pierre Schaeffer. Although a precursor to electronic music as we know it today, it did feature many techniques common today such as changes in pitch, intensity, echo-chamber effects, editing and cut-ups (including time reversal).

FIGURE 12.2: This is the first commercially sold Moog synthesiser produced in 1964. (Public Domain Image.)

12.1.3 The Moog Synthesiser: 1964

The development and release of the Moog synthesiser in 1964 was a key moment in the development of electronic subtractive synthesis (see Figure 12.2).

Bob Moog's seminal paper [2] describes Moog's electronic synthesis innovations in 1964. The significant innovation achieved by Bob Moog was in breaking down subtractive synthesis into a number of functional blocks which could be implemented by standardised modules, e.g., Voltage Con-

trolled Oscillators (VCOs), Voltage Controlled Filters (VCFs), etc. An original
Moog synthesiser is shown in Figure 12.2.

FIGURE 12.3: The inside of a Mellotron. Each tape/playback
head is connected to a key on the keyboard. Obtained from
https://www.flickr.com/photos/44124435310@N01/21520770.
Copyright © Eric Haller, [1] licensed under CC BY 2.0
https://creativecommons.org/licenses/by/2.0/.

12.1.4 The Mellotron: 1974

The Mellotron is an analogue sampler first sold commercially in 1974. It
consists of individual tape loops or strips that had individual recorded musical sounds recorded on them. These were usually at different pitches and
controlled by individual keys on a piano-like keyboard. When a key on the
keyboard was depressed, the associated pre-recorded tape was passed over
a playback head. The effect was that a hugely variable (for the time) number
of sounds could be played simultaneously and harmonically using the keyboard. The sound of the Mellotron can be heard in numerous hit records of
the time. The sound is synonymous in many people's minds with the intro to

---

[1]https://www.flickr.com/photos/haller.

"Strawberry Fields Forever" by the Beatles (one of the first times it was used on a recording). A photo of the inside of a Mellotron is shown in Figure 12.3.

FIGURE 12.4: The DX7 from Yamaha (produced in 1983). (Public Domain Image.)

12.1.5   The DX7 from Yamaha: 1983

The DX7 was the first commercial digital synthesiser produced in 1983. It is one of the best selling synthesisers ever and the first to be based on Frequency Modulation (FM) digital synthesis (see section 12.8). It also had an optional breath sensing controller. A photo of the inside of a Mellotron is shown in Figure 12.4.

12.1.6   MIDI (Musical Instrument Digital Interface): Standardised 1983

MIDI was standardised in 1983, defining not just a communication protocol but electrical connectors and a digital interface for the communication between a wide variety of musical electronic instrument computer interfaces and related controllers. A MIDI link is capable of carrying up to 16 single

FIGURE 12.5: MIDI ports and cables.

MIDI information channels, each of which can be routed to separate devices. A variety of MIDI ports and cables are shown in Figure 12.5.

MIDI is still widely used today for interoperable communication and has been used to control software and hardware-based musical devices using a huge range of possible controlling methods from the commonly found and understood electronic keyboards to more arcane controllers built using hardware tools such as the Arduino.

12.1.7   Cubase Developed for Atari ST

With the introduction of Cubase in 1989 (on the Atari ST platform) a new era of DAW (Digital Audio Workstation) systems was ushered into the modern age of electronic music. Although many competing (and arguably superior) DAW systems have been introduced since the development of Cubase it has been hugely influential in the modern history of electronic music. The Virtual Studio Technology (VST) audio plugin protocol defined initially for Cubase has also been widely adopted. A screenshot of a modern version of cubase is shown in Figure 12.6.

FIGURE 12.6: Cubase 6. [2] Copyright © Steinberg (steinberg.net) licensed under CC BY 3.0 https://creativecommons.org/licenses/by/3.0/.

12.2 Additive Synthesis

Similar to Fourier's original theory, additive musical synthesis is based on the concept that any waveform can be decomposed into a series of periodic functions. Additive synthesis, therefore, forms a musical waveform through the generation and summation of periodic functions. It is the principle used in old-fashioned synthesisers, e.g., the Hammond organ. Figure 12.7 shows the basic principle of additive synthesis, i.e., a summation of a number of oscillators whose frequencies (f_i) and amplitude (r_i) can be controlled individually.

12.3 Temporal Envelopes: ADSR

Each element of the complete output of a musical synthesis can be temporally modified using temporal envelopes. This is usually in the form of a so-called ADSR filter. ADSR stands for Attack, Decay, Sustain and Release.

- Attack: The initial time it takes to get to the maximum value.

- Decay: Once the maximum value is reached the signal decays for a short time.

- Sustain: A steady state is then retained.

---

[2]Source: https://www.steinberg.net/en/landing_pages/c6_creative_commons.

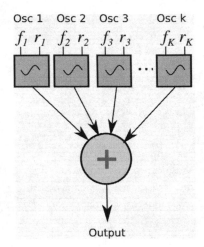

FIGURE 12.7: A basic schematic of additive synthesis.

- Release: After the sustain temporal section this section is defined as the time when the amplitude falls back to zero.

Figure 12.8 shows a basic ADSR temporal filter. This is a simplification of extremely complicated envelopes that can now be generated by modern synthesisers. However, in many cases, ADSR effectively characterises and mimics the temporal amplitude variation of typical musical sounds, e.g., percussive sounds will have a fast attack and decay with a small amount of sustain and release and bowed string sounds have a very long attack and specifically defined decay, sustain and release.

12.4   Subtractive Synthesis

Subtractive synthesis is the process of modulating a generated waveform (or summed waveforms) by filters and envelopes to generate an output waveform. It is a very common form of modern musical synthesis and it is what is commonly known as "electronic synthesis" when referred to as a generic term. Elements of a subtractive analogue synthesiser (see Figure 12.9) are:

- VCO: Voltage Controlled Oscillator (or just OSC).

- VCF: Voltage Controlled Filter.

- VCA: Voltage Controlled Amplifier .

- LFO: Low Frequency Oscillator (used for control)

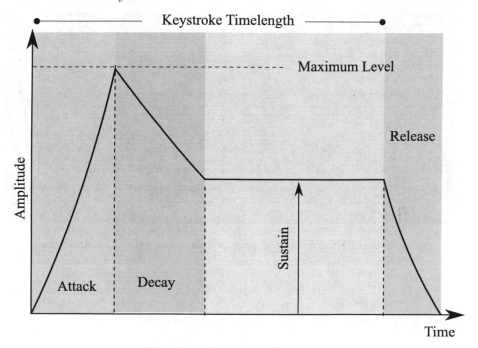

FIGURE 12.8: ADSR temporal filter.

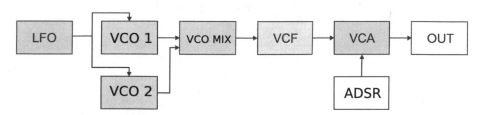

FIGURE 12.9: A typical subtractive synthesis structure diagram.

These elements are shown on the front of the Roland Jupiter JS8 shown in Figure 12.10. This figure also shows an ADSR temporal filter.

12.4.1 Types of Filter for Subtractive Synthesis

The types of filter utilised for subtractive synthesis include lowpass, highpass, bandpass, notch/peak and the comb filter. The first three are illustrated in Figures 12.11, 12.12 and 12.13, respectively. These figures illustrate the key features of these types of filter including the cut-off frequency, resonance amount, the passband, stopband and transition band. The transition bands of the high and lowpass filters are described in terms of the transition slope. This is usually measured in terms of dBs per octave (or decade). For software and

FIGURE 12.10: The Roland Jupiter JP8. (Public Domain Image.)

hardware synthesisers this slope is often characterised as 6,12 or 24 dBs per octave (this can be seen as the HPF (highpass filter) on the cover of Roland's Jupiter 8 hardware synthesiser shown in Figure 12.10).

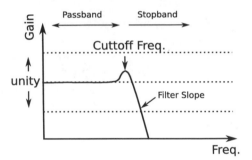

FIGURE 12.11: Low Pass Filter (LPF).

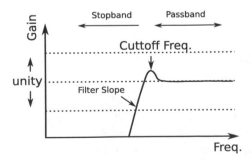

FIGURE 12.12: High Pass Filter (HPF).

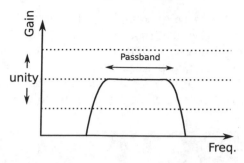

FIGURE 12.13: Band Pass Filter (BPF).

All of these filters can be implemented using simple analogue circuits or IIR/FIR digital filters. The parameters for the filters such as resonance, cutoff frequency, etc. can be adjusted using either variable components in analogue circuits or the calculation of digital filter coefficients using simple filter design formulae.

12.5 Second-Order Digital Filter Design: The Biquad Filter

Although there are numerous methods and architectures for designing and implementing digital filters, the biquad filter is a flexible second-order IIR filter structure that can be used to easily implement tunable lowpass, highpass and bandpass filters whose characteristics (such as cut off frequency) can be tunable in a real-time implementation. An enormous amount of other digital filter methods exist but the biquad will be explained now in detail as it exemplifies many common processes used for real-time filter design.

The biquad is a second-order IIR digital filter and can be represented in the z transform domain as follows:

$$H(z) = \frac{b_0 + b_1 z^{-1} + b_2 z^{-2}}{a_0 + a_1 z^{-1} + a_2 z^{-2}}. \tag{12.1}$$

To reduce the number of parameters by one, the filter coefficients are often normalised as follows (by dividing by a_0):

$$H(z) = \frac{(b_0/a_0) + (b_1/a_0)z^{-1} + (b_2/a_0)z^{-2}}{1 + (a_1/a_0)z^{-1} + (a_2/a_0)z^{-2}}. \tag{12.2}$$

In the time domain, this can be represented as

$$
\begin{aligned}
y[n] = \quad & (b_0/a_0)x[n] + (b_1/a_0)x[n-1] + (b_2/a_0)x[n-2] \\
& -(a_1/a_0)y[n-1] - (a_2/a_0)y[n-2]
\end{aligned}
\tag{12.3}
$$

A circuit diagram of such a biquad is shown in Figure 12.14. In order to define and control the biquad filter, the coefficients a_0, a_1, a_2, b_0, b_1 and b_2 need to be defined/controlled. This can be done in a variety of ways. However, in order to define low, high and bandpass filters a "cookbook" for calculating these parameters from easier to control parameters is often used. A very common "cookbook" for biquad parameters is by Bristow-Johnson [1].

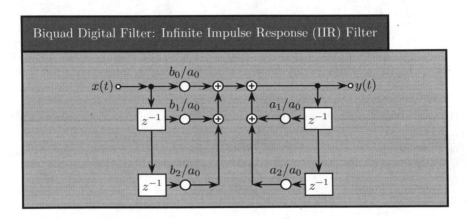

FIGURE 12.14: Biquad digital filter: normalised coefficients.

12.5.1   Implementation of Digital Filters: Biquad Parameter "Cookbook"

The initial stage of the "cookbook" recipe [1] is to define the following

f_s: The sampling frequency.

f_0: Centre frequency (or corner frequency).

Q: Measure of bandwidth.

$\omega_0 = 2\pi f_0/f_s$ (normalised angular frequency).

$\alpha = \sin(\omega_0)/(2Q)$ (intermediate value).

The following tables (Tables 12.1, 12.2 and 12.3) show how three different types of filter can easily be defined and controlled using the Bristow-Johnson cookbook formulae [1]. Furthermore, the following code (Listing 12.1) shows how these filters can be implemented. The following figures (Figures 12.15, 12.16, 12.17, 12.18, 12.19 and 12.20) also show the frequency response of the generated filters together with their representations on the z plane.

TABLE 12.1: Coefficient Calculation for Biquad Filters (LPF and HPFs)

Coeffs	Low Pass Filter (LPF)	High Pass Filter (HPF)
b_0	$(1 - \cos(\omega_0))/2$	$(1 + \cos(\omega_0))/2$
b_1	$1 - \cos(\omega_0)$	$-(1 + \cos(\omega_0))$
b_2	$(1 - \cos(\omega_0))/2$	$(1 + \cos(\omega_0))/2$
a_0	$1 + \alpha$	$1 + \alpha$
a_1	$-2 \times \cos(\omega_0)$	$-2 \times \cos(\omega_0)$
a_2	$1 - \alpha$	$1 - \alpha$

TABLE 12.2: Coefficient Calculation for Biquad Filters (BPF1 and BPF2)

Coeffs	Band Pass Filter 1 (BPF1)	Band Pass Filter 2 (BPF2)
b_0	$\sin(\omega_0)/2$	α
b_1	0	0
b_2	$-\sin(\omega_0)/2$	$-\alpha$
a_0	$1 + \alpha$	$1 + \alpha$
a_1	$-2 \times \cos(\omega_0)$	$-2 \times \cos(\omega_0)$
a_2	$1 - \alpha$	$1 - \alpha$

TABLE 12.3: Coefficient Calculation for Biquad Filters (Notch and APF)

Coeffs	Notch Filter	All Pass Filter (APF)
b_0	1	$1 - \alpha$
b_1	$-2 \times \cos(\omega_0)$	$-2 \times \cos(\omega_0)$
b_2	1	$1 + \alpha$
a_0	$1 + \alpha$	$1 + \alpha$
a_1	$-2 \times \cos(\omega_0)$	$-2 \times \cos(\omega_0)$
a_2	$1 - \alpha$	$1 - \alpha$

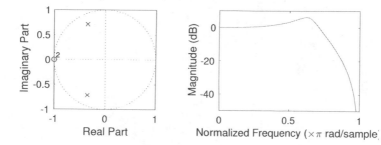

FIGURE 12.15: Lowpass filter.

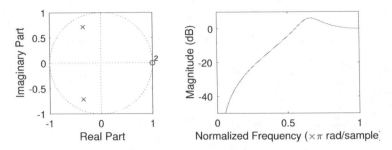

FIGURE 12.16: Highpass filter.

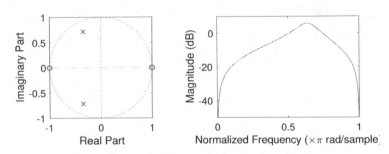

FIGURE 12.17: Bandpass filter 1.

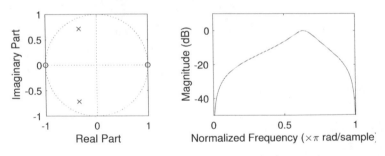

FIGURE 12.18: Bandpass filter 2.

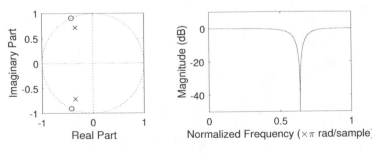

FIGURE 12.19: Notch filter.

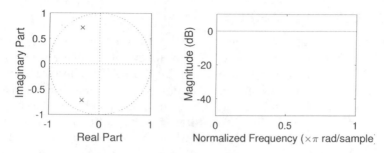

FIGURE 12.20: Allpass filter.

Listing 12.1: Code to implement codebook based biquad filtering

```
1   filtertype = 'lp';
2   % Choose from lp, hp, bp1, bp2, nf or apf.
3   FS = 44100; f0 = 14090; Q =2;
4   w0 = 2*pi*f0/FS;
5   alpha = sin(w0)/(2*Q);
6   switch filtertype
7      case 'lp' %Low pass filter
8         b(1) = (1 - cos(w0))/2; b(2) = 1 - cos(w0); b(3) = (1 -
             cos(w0))/2;
9         a(1) = 1 + alpha; a(2) = -2*cos(w0); a(3) = 1 - alpha;
10     case 'hp' %High pass filter
11        b(1) = (1 + cos(w0))/2; b(2) = -1 - cos(w0); b(3) = (1 +
             cos(w0))/2;
12        a(1) = 1 + alpha; a(2) = -2*cos(w0); a(3) = 1 - alpha;
13     case 'bp1' %Band pass filter 1
14        b(1) = sin(w0)/2; b(2) = 0; b(3) = -sin(w0)/2;
15        a(1) = 1 + alpha; a(2) = -2*cos(w0); a(3) = 1 - alpha;
16     case 'bp2' %Band pass filter 2
17        b(1) = alpha; b(2) = 0; b(3) = -alpha;
18        a(1) = 1 + alpha; a(2) = -2*cos(w0); a(3) = 1 - alpha;
19     case 'nf' %Notch filter
20        b(1) = 1; b(2) = -2*cos(w0); b(3) = 1;
21        a(1) = 1 + alpha; a(2) = -2*cos(w0); a(3) = 1 - alpha;
22     case 'ap' %All pass filter
23        b(1) = 1-alpha; b(2) = -2*cos(w0); b(3) = 1+alpha;
24        a(1) = 1 + alpha; a(2) = -2*cos(w0); a(3) = 1 - alpha;
25  end
26  subplot(1,2,1);
27  zplane(b,a);
28  subplot(1,2,2);
29  [H,W] = freqz(b,a,FS);
30  plot(W/pi,20*log10(abs(H)));
```

12.5.2   Other Digital Filters

A multitude of methods for designing digital filters is possible within (and outside of) MATLAB. Firstly, other biquad formulations have been proposed in the DAFX manual [6] and separately by Zolzer [7]. These references also contain a large number of higher order filters. To get more control of a digital frequency response using biquad filters they can also be used in chains (see below). Other IIR and FIR digital filters can be defined using many other methods similar to the cookbook given above. MATLAB functions fvtool and designfilt can be used together with other tools to precisely define audio digital filters. The MATLAB function dsp.BiquadFilter can be used to design a biquad filter using parameters similar to the cookbook parameters given above.

12.6   Ring Modulation: RM

Ring modulation is the multiplication of one waveform with another. It is a simple way of creating interesting frequency sidebands. Given two spectra S_m (the modulator signal) and S_c (the carrier signal), ring modulation makes a copy of S_m at the location of every frequency in S_c, with each copy scaled by the amplitude of that particular frequency of S_c. Therefore, multiplying two sinusoidal tones produces energy at the sum and difference of the two frequencies. Given the example of a carrier and modulator sinusoid RM can be represented as follows:

$$S_m = \cos(\omega_m t), \tag{12.4}$$
$$S_c = \cos(\omega_c t). \tag{12.5}$$

The multiplication of S_m and S_c is given by:

$$S_{out} = S_m S_c = \cos(\omega_m t)\cos(\omega_c t). \tag{12.6}$$

Given standard identities, S_{out} can be expressed as:

$$S_{out} = 1/2(\cos(\omega_m t - \omega_c t) - \cos(\omega_m t + \omega_c t)), \tag{12.7}$$

where S_m and S_c are the two sinusoids of angular frequencies ω_m and ω_c, respectively and t is the time index. This effect is shown in Figure 12.21. RM often leads to very "inharmonic" synthesis outputs.

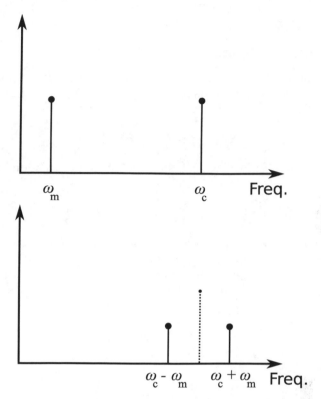

FIGURE 12.21: The ring modulation of two sinusoids at frequency ω_c (the carrier) and ω_m (the modulator).

12.7 Amplitude Modulation: AM

Amplitude modulation (AM) is similar to ring modulation except that firstly it is achieved through modulation of the carrier's amplitude rather than simple multiplication and secondly, the output signal for two sinusoids (carrier and modulator) has the same two output frequencies as ring modulation but also retains the original carrier frequency. This results in a similar result as the bottom figure of Figure 12.21 but with some of original dotted S_c carrier content being retained. Although AM can be more flexibly defined, sinusoidal AM can be represented as:

$$S_{out} = (1 + \alpha S_m)S_c, \qquad (12.8)$$

where S_m and S_c can be defined as in (12.4) and (12.5) and α is the modulation parameter.

12.8   Frequency Modulation: FM

Frequency modulation (FM) is the modulation of the frequency of one signal with another. For two sinusoids (S_m defined above) FM can be expressed as:

$$S_{out} = \cos((\omega_c + S_m)t). \tag{12.9}$$

However this is most usually expressed as "phase modulation":

$$S_{out} = \cos(\omega_c t + \beta \sin(\omega_m t)). \tag{12.10}$$

For a sinusoid carrier of frequency ω_c and modulator of frequency ω_m amplitude and ring modulation generates just two sidebands at frequencies $(\omega_c + \omega_m)$ and $(\omega_c - \omega_m)$. FM generates a series of sidebands (the position of which is illustrated in Figure 12.22):

$$\omega_{sb} = \omega_c \pm n\omega_m \quad \text{where } n \in \mathbb{N}. \tag{12.11}$$

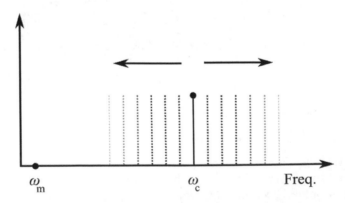

FIGURE 12.22: Positions of the sidebands of FM synthesis.

The weight of each sideband is defined by the "modulation index" β defined in (12.10). β is also defined as the ratio of the maximum change in the carrier frequency modulation to the frequency of the modulation signal, i.e.,

$$\beta = \frac{\Delta\omega_c}{\omega_m} \tag{12.12}$$

Frequency modulation can be expressed as an infinite sum of sidebands where the weights of each sinusoid are given by the Bessel function of the first kind $J_n(\beta)$:

$$\cos(\omega_c t + \beta \sin(\omega_m t)) = \sum_{n=-\infty}^{\infty} J_n(\beta)\cos((\omega_c + n\omega_m)t). \qquad (12.13)$$

Equation (12.13) is illustrated in both Figures 12.22 and 12.23. Figure 12.22 shows the position of the sidebands and Figure 12.23 shows the weights of all the sidebands given by the Bessel function of the first kind $J_n(\beta)$. The modulation index β therefore controls the weights of the sidebands with larger values of β leading to larger weights for the sidebands further from the carrier (i.e., $\beta=0$ leads to an output of just the carrier). Figure 12.23 is created by the following MATLAB code.

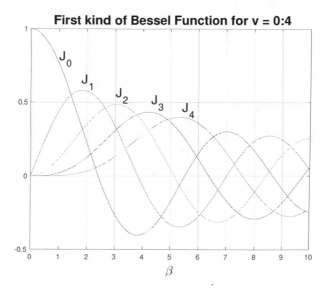

FIGURE 12.23: Bessel functions of the first kind for FM weights of Equation (12.13).

```
1  clear all;
2  beta = 0:0.1:10;
3  for indx = 1:5
4      J(indx,:) = besselj(indx-1,beta);
5  end
6  plot(beta,J)
7  text(1,0.8,'J_0'); text(1.8,0.64,'J_1'); text(3,0.55,'J_2');
       text(4.3,0.5,'J_3'); text(5.4,0.45,'J_4');
8  grid on
```

```
9  title('First kind of Bessel Function for v = 0:4', 'fontsize',
       18);
10 xlabel('\beta', 'fontsize', 18);
```

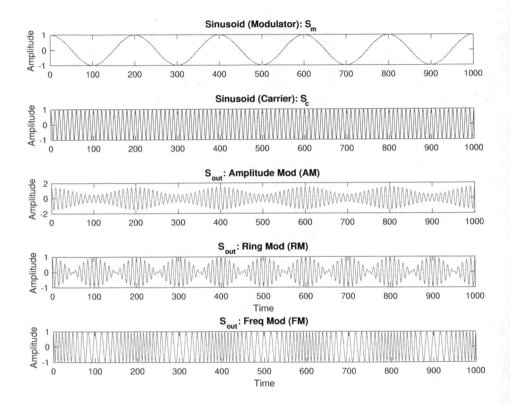

FIGURE 12.24: Comparison of ring modulation (RM), amplitude modulation (AM) and frequency modulation (FM).

Frequency modulation can be achieved with arbitrary carrier and modulation signals to give complex harmonic content. A comparison of AM, RM and FM is shown in Figure 12.24 which was created by code in Listing 12.2.

Listing 12.2: Comparison of Modulation Methods

```
1   wm = 0.005; wc = 0.1; m = 0.5; beta = 8;
2   t=0:1:1000;
3   Sm=cos(2*pi*wm*t);
4   Sc=cos(2*pi*wc*t);
5   m = 0.5;
6   AM=(1+m.*Sm).*Sc;
7   RM= Sm.*Sc;
8   FM = cos(2*pi*wc*t+beta.*sin(2*pi*wm*t));
9   subplot(511);
10  plot(t,Sm);
11
12  ylabel('Amplitude');
13  title('Sinusoid (Modulator): S_m')
14  subplot(512)
15  plot(t,Sc);
16  ylabel('Amplitude');
17  title('Sinusoid (Carrier): S_c');
18  subplot(513);
19  plot(t,AM);
20  plot(t,AM,'r');
21  ylabel('Amplitude');
22  title('S_{out}: Amplitude Mod (AM)');
23  subplot(514);
24  plot(t,RM);
25  plot(t,RM,'r');
26  xlabel('Time');
27  ylabel('Amplitude');
28  title('S_{out}: Ring Mod (RM)');
29  subplot(515);
30  plot(t,FM,'r');
31  xlabel('Time');
32  ylabel('Amplitude');
33  title('S_{out}: Freq Mod (FM)');
```

12.8.1 Frequency Modulation/Biquad Filtering: Matlab Example Implementation

Figure 12.25 shows the output spectrogram of the FM test code given in Listing 12.3. This code is an example implementation of a frequency modulation to create a bell-like sound. This is achieved by making the modulation sinusoid frequency have an inharmonic relationship to the frequency of the carrier sinusoid frequency. As is common, the FM modulation index is also modulated over time. Finally, the entire sound amplitude is modulated using a simple exponential function. The spectrogram of the output shown in Figure 12.25 shows the central carrier frequency and the sidebands decreasing in number and strength over time as the modulation index β is decreased by the exponential function over time. This gives a surprisingly realistic bell-like sound for such a small amount of code.

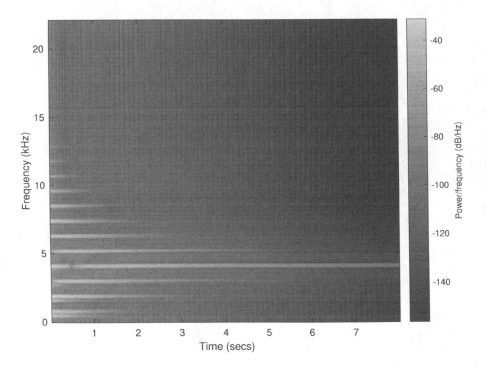

FIGURE 12.25: Frequency modulation example spectrogram output (from code in Listing 12.3).

Listing 12.3: FM Test Code

```
1  Fs = 44100; Fc = 4000; Fm = 1100;
2  Duration = 8;
3  t=0:1/Fs:Duration;
4
5  A=exp(-t);
6  B=4*exp(-t);
7
8  %FM Synthesis
```

```
9   x = A.*cos(2*pi*Fc*t+B.*cos(2*pi*Fm*t));
10
11  soundsc(x, Fs);
12  spectrogram(x, 512, 256, [], Fs, 'yaxis');
```

Figure 12.26 shows the output spectrogram of the FM and biquad test code given in Listing 12.4. This is basically the same code as shown in Listing 12.3 but with the output of the FM synthesis being filtered by a specifically designed biquad LPF. As is common, this biquad LPF is used to filter twice in a chain to give the final output. The spectrogram shown in Figure 12.26 shows how the higher frequencies are attenuated when compared to the previous spectrogram shown in Figure 12.25.

A similar code and output is shown in Figure 12.27 and Listing 12.5. However, this figure and code show the subtractive synthesis system using a square wave filtered by a biquad filter and an amplitude modulation. This exemplifies a simple subtractive synthesis.

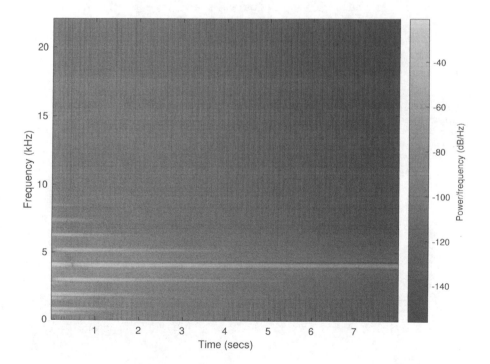

FIGURE 12.26: FM and biquad example spectrogram output (from code in Listing 12.4).

Listing 12.4: Combined FM and Biquad Synthesis

```
1  Fs = 44100; Fc = 4000; Fm = 1100;
2  Duration = 8;
3  t=0:1/Fs:Duration;
4
5  A=exp(-t); I=4*exp(-t);
6
7  %FM Synthesis
8  x = A.*cos(2*pi*Fc*t+I.*cos(2*pi*Fm*t));
9
10 %Biquad LPF
11 f0 = 4090; Q =2;
12 w0 = 2*pi*f0/Fs;
13 alpha = sin(w0)/(2*Q);
14
15 %Coeffs from cookbook
16 b(1) = (1 - cos(w0))/2; b(2) = 1 - cos(w0); b(3) = (1 - cos(w0))
      /2;
17 a(1) = 1 + alpha; a(2) = -2*cos(w0); a(3) = 1 - alpha;
18
19 %Filter FM output twice with Biquad
20 x = filter(b,a,x);
21 x = filter(b,a,x);
22
23 soundsc(x, Fs);
24 spectrogram(x, 512, 256, [], Fs, 'yaxis');
```

Listing 12.5: Combined Square Wave and Biquad Synthesis

```
1  Fs = 44100; Fc = 4000; Fm = 1100;
2  Duration = 8;
3  t=0:1/Fs:Duration;
4
5  A=exp(-t); I=4*exp(-t);
6
7  %Square wave
8  x = A.*square(2*pi*Fc*t);
9
10 %Biquad LPF
11 f0 = 4090; Q =2;
12 w0 = 2*pi*f0/Fs;
13 alpha = sin(w0)/(2*Q);
14
15 %Coeffs from cookbook
```

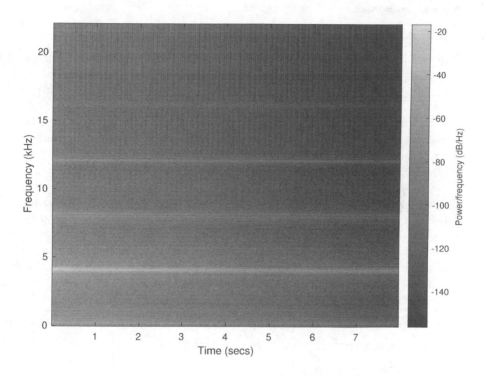

FIGURE 12.27: Square wave and biquad example spectrogram output (from code in Listing 12.5).

```
16  b(1) = (1 - cos(w0))/2; b(2) = 1 - cos(w0); b(3) = (1 - cos(w0))
        /2;
17  a(1) = 1 + alpha; a(2) = -2*cos(w0); a(3) = 1 - alpha;
18
19  %Filter FM output with Biquad
20  x = filter(b,a,x);
21
22  soundsc(x, Fs);
23  spectrogram(x, 512, 256, [], Fs, 'yaxis');
```

12.9   Alternative Musical Synthesis Techniques

Wavetable synthesis is very similar to subtractive synthesis. However, instead of using a simple oscillator such as a square wave or sinusoid, wavetable synthesis uses a lookup table of values that are repeatedly output at a given amplitude and frequency. It is an efficient method that avoids complex calculations for each sample. The value at the current position in the lookup table is output at a point in time. Interpolation between adjacent data points can achieve greater precision. Use of amplitude, filter and low-frequency modulation techniques borrowed from subtractive synthesis can be applied to the repeated table. Granular synthesis replays short audio sections (known as grains) sequentially. These grains are no more than 50ms long and typically 20–30ms. They are most often derived from an entire pre-existing sample (which is segmented into overlapping grains). They can also be generated by another form of synthesis (wavetable, FM, etc). The grains can be replayed at any chosen rate and transformed using an envelope together typically with pitch and amplitude shifting.

12.10   Time Warping

Time warping can be achieved using a variety of techniques. Time domain techniques use the overlap and add structure discussed in Chapter 7. Variants include

SOLA: Synchronised Overlap and Add [4].

WSOLA: Wave Similarity Overlap and Add [5].

PSOLA: Pitch Synchronous Overlap and Add [3].

Another common method of time warping is achieved in the frequency domain. This is most often implemented using the so-called Phase Vocoder.

12.10.1   Time Warping Using the Phase Vocoder

A common mistake in implementing a time warping method in the frequency domain is to simply modify the rate of the output grains within a time-frequency modification code. However, the changes in the phases for the new rate will not be correct in terms of steady-state tones. This issue is illustrated in Figure 12.28 which displays the spectrogram and phasogram of some steady state frequencies. The phase changes (in the x directions of the phasogram) at the different input frequencies in relation to the frequency (in the y direction). Retaining the same variation in phase changes from the input will not, therefore, result in effective time warping. Time warping within the frequency domain therefore requires:

- The playback of the reconstructed grains at the new rate

- The adjustment of the phases with each grain (in the transform domain)

- Adjustment of the output gain (discussed in Chapter 7).

A detailed description of phase modification for time warping is given by Zolzer in the DAFX manual [7].

Modification of the phase of an audio signal can give interesting audio effects. Setting the phase to zero can give an interesting "robot" effect for speech as demonstrated in Listing 12.6. This code also illustrates a different application of time-frequency processing using FFTs, inverse FFTs, weighted analysis and weighted synthesis windows (using the COLA condition) as described previously in Chapter 3.

Listing 12.6: Phase Vocoder for Robot Effect

```
1  [xIn,Fs] = audioread('paulhill2.wav');
2
3  L = length(xIn);
4  fshift = 128;
5  lenW = 2048;
6  window1 = hann(lenW);
7  ratioCOLA =sum(window1.^2)/fshift;
8  window1 =window1/sqrt(ratioCOLA);
9  window2 = window1;
10
11  xIn = [zeros(lenW, 1); xIn; zeros(lenW-mod(L,fshift),1)];
12  xOut = zeros(length(xIn),1);
13  pntrIn = 0; pntrOut = 0;
14
15  while pntrIn<length(xIn) - lenW
16         thisGrain = window1.*xIn(pntrIn+1:pntrIn+lenW);
17         f = fftshift(fft(thisGrain));
18         magF = abs(f);
19         phaseF = 0*angle(f); % Set phase to zero ("robot effect")
           ;
20         fHat = magF.* exp(i*phaseF);
21         thisGrain = window2.*real(ifft(fftshift(fHat)));
22
23         xOut(pntrOut+1:pntrOut+lenW) = thisGrain + xOut(pntrOut
               +1:pntrOut+lenW);
24         pntrOut = pntrOut + fshift;
```

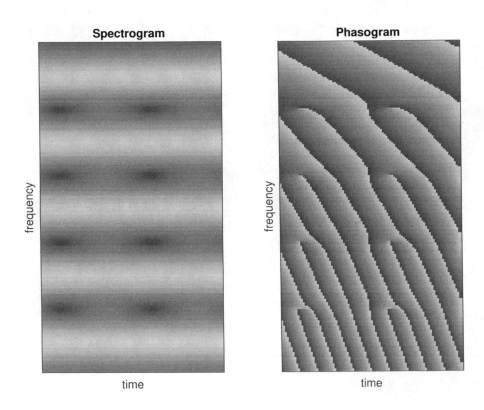

FIGURE 12.28: Spectrogram and phasogram of steady state frequency content.

```
25    pntrIn = pntrIn + fshift;
26 end
27
28 soundsc(xOut, Fs);
```

12.11 Musical Synthesis: Summary

- Four main types of synthesis.

  - Additive synthesis.
  - Subtractive synthesis.
  - Wavetable synthesis.
  - Granular synthesis.

- Different types of modulation produce differing effects.

  - Amplitude Modulation (AM).
  - Ring Modulation (RM).
  - Frequency Modulation (FM).

- Phase vocoder is for time-frequency manipulation.

  - Phase manipulation for time-stretching/shrinking.
  - Phase set to 0 for "robot" effect.
  - Random phase for "chorus" effect.

12.12 Exercises

Exercise 12.1
Modify the code in Listing 12.6 to produce random phase in the output. Does this output sound like a "chorus" effect?

Exercise 12.2
Modify the code in Listing 12.4 to produce more of a "wind instrument sound."

Exercise 12.3
Describe the key difference in the output of Ring and Amplitude Modulation when the two signals (carrier and modulator) are sinusoids.

Bibliography

[1] Cookbook formulae for audio EQ biquad filter coefficients
`http://www.musicdsp.org/files/Audio-EQ-Cookbook.txt`, (accessed
February 3rd, 2018).

[2] R.A. Moog. Voltage-controlled electronic music modules. *Journal of the
Audio Engineering Society*, 13(3):200–206, 1965.

[3] E. Moulines and F. Charpentier. Pitch-synchronous waveform processing
techniques for text-to-speech synthesis using diphones. *Speech Communication*, 9(5-6):453–467, 1990.

[4] S. Roucos and A. Wilgus. High quality time-scale modification for speech.
In *Acoustics, Speech, and Signal Processing, IEEE International Conference on
ICASSP'85.*, volume 10, pages 493–496. IEEE, 1985.

[5] W. Verhelst and M. Roelands. An overlap-add technique based on waveform similarity (wsola) for high quality time-scale modification of speech.
In *Acoustics, Speech, and Signal Processing, 1993. ICASSP-93., 1993 IEEE International Conference on*, volume 2, pages 554–557. IEEE, 1993.

[6] U. Zölzer. *Digital Audio Signal Processing*. John Wiley & Sons, 2008.

[7] U. Zölzer. *DAFX: Digital Audio Effects*. John Wiley & Sons, 2011.

Appendices

A

The Initial History of Complex Numbers

CONTENTS

A.1 Cardano's Formula

Although many previous mathematicians contributed to their history, Gerolamo Cardano (or Cardan) is recognised as initiating the effective use of complex numbers. However, their history around the date of their discovery is somewhat more complicated. Scipione del Ferro (d. 1526) was the first to be able to algebraically solve a simplified cubic equation around 1515 in Bologna, northern Italy. No manuscripts of his method exist. However, through several suggested routes, the method became known and illustrated by Cardano, culminating in its publication in 1545 within his Ars Magna, the first Latin treatise on algebra.

This solution to cubic equations became intricately linked to the development of complex numbers. Firstly, the cubic solved was a slightly simplified version of a general cubic known as the depressed form (i.e., without the square term): [1]

$$x^3 + bx + c = 0 \tag{A.1}$$

The solution to this equation was found to be

$$x = \sqrt[3]{-\frac{c}{2} + \sqrt{\frac{c^2}{4} - \frac{b^3}{27}}} + \sqrt[3]{-\frac{c}{2} - \sqrt{\frac{c^2}{4} - \frac{b^3}{27}}}. \tag{A.2}$$

This equation solved the depressed cubic Equation (A.1) in many cases. However, the route to the solution is complicated for cases such as $b = -15$ and $c = -4$ (giving the solution using (A.2) as)

$$x = \sqrt[3]{2 + \sqrt{-121}} + \sqrt[3]{2 - \sqrt{-121}}. \tag{A.3}$$

---

[1]It can be shown that a generic cubic equation can be easily transformed into the form shown in (A.1).

317

It can be easily checked that the solution to this equation is 4. However, (A.3) involves the square root of a negative number and therefore at that time not directly solvable.

It is first guessed that the two terms in (A.3) were of the form $a + in$ and $a - in$. As their sum is 4, a must equal 2. Therefore what is required is

$$2 + in = \sqrt[3]{2 + 11i} \qquad\qquad\qquad (A.4)$$

and

$$2 - in = \sqrt[3]{2 - 11i} \qquad\qquad\qquad (A.5)$$

Considering first (A.4): If we take the cube of both sides we get $(2 + in) \times (2 + in) \times (2 + in) = (4 + 4in - n^2)(2 + in) = 8 + 12in - 2n^2 - 4n^2 - in^3$.

Taking the cube of the left hand side we have $8 + 12in - 6n^2 - in^3$. If $n = 1$ $(2 + in)^3 = 2 + 11i$ which is the cube of the right-hand side. Following a similar argument for (A.5) the following is obtained:

$$\sqrt[3]{2 + 11i} + \sqrt[3]{2 - 11i} = (2 + i) + (2 - i) = 4. \qquad\qquad (A.6)$$

A real number solution is obtained through the use of complex numbers. In fact it would be very difficult to solve this case without the use of complex numbers.

This solution gave great impetuous to the development of complex numbers as in this case a real number problem has a real number solution but the solution is found using complex arithmetic.

B

MATLAB *Fundamentals (Applicable to Audio Processing)*

CONTENTS

B.1 Introduction

MATLAB is a numerical computing environment integrated with a matrix-based interpreted programming language. This language allows easy manipulation of matrices, signal processing algorithm development, plotting of data, integration of domain-specific toolboxes together with many other capabilities. MATLAB is both a programming language and an Integrated Development Environment (IDE). The MATLAB IDE consists of:

- A command line interface.

  - Simple commands (including debugging commands) can be evaluated one line at a time.

- A file editor.

- An integrated debugging system.

  - Scripts and functions can be interactively debugged together.

- Visualisation tools.

- A local directory navigation system.

- A workspace showing defined variables and their contents.

Although key to the use of MATLAB, a description of the usage and capabilities of the MATLAB IDE is left to the innumerable tutorials, help files and demos provided by Mathworks and within MATLAB itself. This appendix instead focuses on the MATLAB core of the programming language.

Although the MATLAB language is an interpreted language (i.e. it is not compiled but directly interpreted at runtime), collections of functions can be implemented within the same or different files to form complex interacting collections of source code in the same way as compiled languages. MATLAB has been used for many other signal processing domains, notably image and video processing, control programming and low-level DSP signal processing.

A key advantage of MATLAB is that high-level matrix, linear algebra and signal manipulation tools are built into the system and can be used immediately from the command line or through the running of scripts and/or functions using the MATLAB programming language. The language elements are not tightly defined (weakly typed) and MATLAB is therefore ideally suited to prototyping small to medium sized code.

It is specifically suited to the prototyping of audio processing and manipulation algorithms due to this large array of matrix and signal processing capabilities together with specific audio handling functions. The following tables give a brief description of the key functions, operators and commands of the MATLAB language and command line.

B.1.1 Variables and their Assignment

At the core of MATLAB is the concept of the variable. Variables represent entities within the MATLAB environment in a similar way to variables in other languages. They can be assigned using the assignment operator = and their name must adhere to a small list of conditions.

- Their name must begin with an alphabetic letter.

- Their name is case sensitive.

- The name of the variable must be less than a given length. [1]

- They must not be the same as any of MATLAB's built-in **keywords** or **reserved words**.

- It is problematic if they have the same name as any built-in or user function. This should be avoided to prevent confusion.

- You do not have to declare variables or their type before assigning values to them.

B.1.2 Key MATLAB® Representations and Manipulations

Variable Types (Classes): Each variable can have a different type similarly defined to the types used in languages such as C, C++ or Java. Types are

---

[1]You can find this length using the. `namelengthmax` function (it's 63 on my machine).

known as classes in MATLAB. Common variable classes in MATLAB include double, integer (either signed or unsigned with 16,32 or 64 bits), char and logical.[2]

Vector and Matrix Representations: Variable representations can be scalars, vectors or matrices. This is unlike languages such as C, C++ or Java which have to use (sometimes complicated) extensions in order to represent vectors and matrices.

Native Complex Number Representation: Due to the importance of complex number representations within signal processing, any variable (or its elements if a matrix or vector) can be a complex number. For example, initialising a complex number a = 1+i*4; or initialising a complex vector A = [1+i*4 2+i*3 3*i*2];. The square root of -1 is represented by either i (given the convention used by mathematicians) or j (given the convention used by engineers).

Ordinal Indexing: The indexing of an array or matrix within MATLAB always starts with 1 (see Box 2.5.1).

Transformations and Transpositions: The transformation of an array A or matrix M is A' and M', respectively. The transposition of a row vector v to a column vector is v'. Similarly, the transposition of a column vector v to a row vector is v'.

Simple Arithmetic of Scalars, Vectors and Matrices: Defining a scalar, vector or matrix a, a scalar can be added or subtracted through the simple use of the + or - operators. To add or subtract two matrices (or vectors) element wise these exact same operators can be used.

Matrix and Element-Wise Simple Arithmetic of Arrays and Matrices: Multiplication of two matrices or vectors is achieved for common multiplication rules using the * operator. Similarly, the division operator / divides one matrix by the other by taking the inverse of the second and multiplying it by the first. Simple arithmetic element by element can be achieved by pre-prending the operator (e.g., *, ^, /) by a period (e.g., .*, .^, ./).

Intialisation: The initialisation of vectors and multidimensional indices using function calls such as ones, zeros, rand, etc. uses the same form, e.g., a = ones(thisSize); where thisSize is [10,1] for a vector of length 10 and [10,20,30] is a 3D matrix of dimension (10,20,30).

Vector and Matrix Subsets: Subsets of vectors and / or matrices can be indexed using usual matrix indices, i.e., A(20:30) or A(:,20:40) or A(10:end,20:end); or A(:,20:end);.

Audio signals are often represented as real valued vectors (arrays) of class double.

---

[2]The logical class is MATLAB's boolean type.

B.2 Commands, Assignments, Operators, etc.

Tables B.1, B.2, B.3, B.5 and B.4 show MATLAB's basic commands, assignments, operators, basic arithmetic and plotting commands respectively.

TABLE B.1: Basics Commands

Command	Purpose
;	Indicates the end of a line in a function or command (prevents viewing command output)
%	The start of a comment
!	The start of a system command
clc	Clears the command line
clear a, b, c	Clears the variables a, b and c from the workspace
clear all	Clears the workspace of all variables
close	Closes current figure
close(h)	Closes current figure with handle h
close all	Closes all open figures
save <filename>	Saves all variables in the workspace to <filename>
save <filename> a b c	Saves workspace variables a, b and c to <filename>
load <filename>	Loads saved variables into current workspace
gca	Handle to current axis
gcf	Handle to current figure
gco	Handle to current object
path	Outputs the current path
addpath <path>	Adds <path> to function search path
rmpath <path>	Remove <path> from function search path
fprintf(...	Format the output of variables in a similar way to C++/C printf and fprintf functions
whos	List all currently defined variables
global a b c	Make variables a, b and c have global scope
help <functionname>	Gives help (if it exists) for <functionname>
doc <functionname>	Outputs documentation (if it exists) for <functionname>

TABLE B.2: Defining Scalar, Vector and Matrix Variables

Command	Purpose
a = 10	define and initialise scalar variable a to have value 10
v = [1 2 3]	define and initialise vector v to be the row vector $(1, 2, 3)$
v = [1; 2; 3]	define and initialise vector v to be the column vector $(1, 2, 3)^T$
m = [1 2 3; 4 5 6; 7 8 9]	define and initialise matrix m to be the matrix $\begin{bmatrix} 1 & 2 & 3 \\ 4 & 5 & 6 \\ 7 & 8 & 9 \end{bmatrix}$
v=linspace(x1, x2, n)	define and initialise row vector v of n linearly equally spaced points between x1 and x2.
m=eye(n)	define and initialise an $n \times n$ identiy matrix m.
x2=repmat(x1,m,n)	Define x2 as an $m \times n$ where each element is x1.
x3=[x1 x2];	Concatenates matrices x1 ($m \times n$) and x2 ($m \times k$). x2 adds additional columns to x1.
x3=[x1;x2];	Concatenates matrices x1 ($m \times n$) and x2 ($k \times n$). x2 adds additional rows to x1.

TABLE B.4: Plotting Commands

Command	Purpose
figure	Create a new figure window
plot(y)	Plot the points in array y (y-axis) with x-axis having ordinal index (1,2,3,...)
plot(x,y)	For equal length arrays x axis point positions given by x and y axis point positions given by y
semilogx(x,y)	Same as plot(x,y) but x values are plotted as log(x)
semilogy(x,y)	Same as plot(x,y) but y values are plotted as log(y)
loglogx(x,y)	Same as plot(x,y) but x and y values are plotted as log(x) and log(y) respectively
title('text')	Set the title of the plot to 'text'
xlabel('text')	Label the x axis with 'text'
ylabel('text')	Label the y axis with 'text'
axis equal	Sets the aspect ratio so that tick mark increments on the x and y axes are equal in size
grid	Add a grid to the plot

| hold on | Any new plot will plotted on any existing and current figure |
| hold off | Overwrite old plot by any new plot |

TABLE B.5: Other Basics (Artithmetic)

Command	Purpose
i or j	$\sqrt{-1}$
real(A)	Output the real components of each of the elements of A
imag(A)	Output the imaginary components of each of the elements of A
abs(A)	Output the magnitude of real or complex components of each of the elements of A
angle(A)	Output the phase angle of complex components of each of the elements of A
inv(A)	Take inverse of matrix A
rand(N)	Create uniform distributed random number matrix $(N \times N)$ in the range 0 to 1
randn(N)	Create normally distributed random number matrix $(N \times N)$
rand(N,M)	Create uniform distributed random number matrix $(N \times M)$ in the range 0 to 1. Set either N or M to 1 to create a vector
randn(N,M)	Create normally distributed random number matrix $(N \times M)$. Set either N or M to 1 to create a vector
zeros(N)	Create square matrix where all elements are 0 $(N \times N)$
zeros(N,M)	Create matrix where all elements are 0 $(N \times M)$. Set either N or M to 1 to create a vector
length(A)	Returns the length of vector A
size(A)	Returns size of the multidimensional matrix A
round(A)	Round the elements of A towards nearest integer
floor(A)	Round down the elements of A towards $-\infty$
ceil(A)	Round up the elements of A towards ∞
fix(A)	Round up the elements of A towards 0
sum(A)	If A is a vector sum(A) is the sum of its elements. If A is matrix, sum(A) is a row vector whose elements are the column sums of A
mean(A)	Operates as with sum(A) but with mean output
min(A)	Operates as with sum(A) but with minimum output
max(A)	Operates as with sum(A) but with maximum output
std(A)	Operates as with sum(A) but with standard deviation output

TABLE B.3: Operators

Operator	Example	Description	Type
=	s = 5;	Assigns the value on the right to the variable on the left	Assignment
+	s = t + u;	Adds each element of t to each element of u. If u is a scalar then it is added to each element of t	Arithmetic
−	s = t - u;	Subtracts each element of u from each element of t. If u is a scalar then it is subtracted to each element of t	Arithmetic
^	a = b^c;	Raises b to the power of c.	Arithmetic
.^	A = B.^C;	As with ^ but element by element	Arithmetic
*	A = B * C;	Multiplies B by C. If B by C are both matrices/vectors, then this is done by standard matrix/vector rules.	Arithmetic
.*	A = B.* C;	As with * but element by element	Arithmetic
/	A = B / C;	If C has a single element then the output is each element of B divided by C. If C is a matrix then the output is the inverse of C multiplied by B using standard matrix/vector rules.	Arithmetic
==	if (a==b)	Equal to	Relational operator
~=	if (a~=b)	Not equal to	Relational operator
&&	(a&&b)	Logical AND	Logical operator
||	(a||b)	Logical OR	Logical operator
~	(~b)	Logical NOT	Logical operator

Index